Serving Louisiana

Favorite Recipes of Family and Friends of the LSU AgCenter

Serving Louisiana

Favorite Recipes of Family and Friends of the LSU AgCenter
Published by LSU Agricultural Center

Copyright 2002 by
LSU Agricultural Center
P.O. Box 25203
Baton Rouge, Louisiana 70894-5203

This cookbook is a collection of favorite recipes, which are not necessarily original recipes.

Library of Congress Catalog Number: 2001132616
ISBN: 0-9649132-0-8

Edited, Designed, and Manufactured by
Favorite Recipes® Press
An imprint of

FRP™

P.O. Box 305142
Nashville, Tennessee 37230
(800) 358-0560

Art Director: Steve Newman
Book Design: Malone Creative
Project Manager: Debbie Van Mol
Photography: Donn Young, Donn Young Photography

Manufactured in the United States of America
First Printing: 2002
7,500 copies

About the cover:

The cover of this cookbook was photographed on the grounds of the LSU Rural Life Museum, an historical jewel surrounded by the city of Baton Rouge. The Rural Life Museum is located on the 450-acre site of the LSU AgCenter's Burden Center. The museum provides insight into the largely forgotten lifestyles and cultures of pre-industrial Louisiana. Home to an extensive collection of tools, household utensils, furniture, vehicles, and farming implements, the museum preserves an important part of the rural heritage of Louisiana and the nation. The museum includes more than twenty buildings, spread over five acres of the former plantation.

Pictured on the cover is an old farm wagon with an array of Louisiana food products...many of which are contained in recipes in this cookbook. In the background is an 1800s vintage pioneer cabin.

Photographed by Donn Young, Donn Young Photography, New Orleans, Louisiana

The LSU AgCenter dedicates this book to

Ganelle S. Bullock

in appreciation for her many years of leadership,

service, and commitment to the AgCenter.

Her energy and enthusiasm in the publishing of

this cookbook have been an inspiration and

have truly made this project possible.

11

29

47

65

99

133

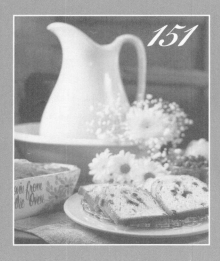

151

169

203

Serving Louisiana is the perfect title for our cookbook. Its pages contain not only delicious recipes served in Louisiana, but valuable facts about how the LSU AgCenter is "serving Louisiana."

Most Americans expect only the highest quality food—and plenty of it—when they go to the grocery store. They also expect the food to be reasonably priced and safe for their families. The LSU AgCenter works every day to make sure Louisianians get the food they expect...and more.

Our research and education programs supply critical information and technology for the state's multibillion dollar agricultural industry, from suppliers to producers to processors. We also provide important research and education programs in human health and nutrition.

Scientists working in the AgCenter's departments on the Baton Rouge campus and twenty research stations throughout the state have developed new and better foods. And they are developing products and systems that help farmers and processors produce and deliver that food more efficiently.

You may recognize the AgCenter's Extension office in your parish as the home of your county agent, home economist, and 4-H agent. These LSU AgCenter faculty members conduct educational programs, visit homes, answer questions, provide hundreds of publications and newsletters, and conduct other activities and events—all to provide the latest information to families in Louisiana's sixty-four parishes. The LSU AgCenter may be best known for working with young people through the 4-H youth development program. The 4-H program focuses on teaching skills that will help young people the rest of their lives.

The LSU AgCenter extends its efforts beyond the borders of Louisiana and the nation. The AgCenter's international programs maintain contacts with agricultural research counterparts in countries around the world to learn how agricultural production and marketing in other parts of the world can benefit Louisiana's agricultural industry.

So, what is the LSU AgCenter? It's Louisiana's unique campus of higher education dedicated to strengthening agriculture and the food system, supporting families, training leaders, and teaching young people. The state is our campus. The LSU AgCenter is "serving Louisiana."

William B. Richardson

William B. Richardson
Chancellor and Chalkley Family Endowed Chair

Welcome to *Serving Louisiana*, the LSU AgCenter's serving of a great tasting collection of recipes from the heart and soul of Louisiana. From the basics (how to make a roux) to local favorites (Blackened Fish Fillets, Louisiana Baked Oysters, Crawfish Étouffée, and New Orleans Chicken), we tell you everything you need to know to cook the Louisiana way—with flavor, gusto, and good friends.

Louisiana's coastline and bayous produce an abundance of seafood. Shellfish and finfish from freshwater and saltwater provide the main ingredients for a variety of Creole and Cajun dishes. And where would those dishes be without rice—a Louisiana staple along with the yams that find their way to Louisiana tables every year?

Like many subtropical and tropical parts of the world, Louisiana produces an array of herbs and spices—particularly the pungent filé that seasons gumbos and spicy peppers, whose sauces enhance all manner of Louisiana cuisine.

Louisianians have learned how to combine local products into interesting foods. Alligator and venison sausages are widely used in jambalaya. Sweet cane sugar and pecans combine to become mouthwatering pralines. And andouille and tasso can be found in everything from the simplest country fare to gourmet dishes in New Orleans' finest restaurants.

The recipes in this cookbook are favorite recipes of our AgCenter family and friends and come from all over the state—some traditional and some quick-and-easy, but all guaranteed to be Louisiana's best.

You'll find favorite recipes for every part of the meal, from rolls hot from the oven (Refrigerator Rolls and Whole Wheat Refrigerator Rolls) to sweet treats that will make your sweet tooth ache (Creamy Pralines, Italian Cream Cake, and Sour Cream Pound Cake). There's even a special chapter titled "Lagniappe" (a little something extra) that talks about proper table settings, how to prepare food for fifty, and other important food facts.

So as we serve up an array of scrumptious recipes, we hope you'll delight in *Serving Louisiana*. Take it home, put on your apron, and be prepared to enjoy!

Michele Abington-Cooper

Connie Aclin

Kristy Addison

Mike Anderson

Vivian & Ronnie Anderson

Winzer Andrews

Linda Arneson

Betty Bagent

Cleo Barber

Gale Bateman

Mary Virginia Benhard

Margaret Ann Blackwell

Karla Blalock

Pat Blanchard

Betty Boethel

John Brady

Dorothy Britton

Barbara Brown

Beverly Bruce

Ganelle Bullock

Lance Bullock

Martha Burch

Rouse Caffey

Becky Calhoun

Ann Callegari

Sheryl Carnegie

Christy Carriere

Kay Collins

Leaudry Connor

Lillian Conway

Arlene & Paul Coreil

Ann Coulon

Norma Culkin

Jennifer Dauzat

Julie Dauzat

Laura & Ruben Dauzat

Tommy Dauzat

Eunice Davis

Jo Anne & Wayne Davis

Cornelis deHoop

Sharon Dixon

Heather Doles

Polly Doles

Peggy Draughn

Vicki Dunaway

Delaine & Mark Emmert

Mary Faulkner

Lee Ann Fields

Chef John Folse

Kandice Fontenot

Dale Frederick

Eloise Futrell

Lyda Gatewood

Gayle & Byron Gautreau

Toni Gilboy

Danna Gillett

Yvette Girouard

Joan Gobert

Minus Granger

Lillian Green

Ann Guedry

Shannon Guglielmo

Blanche Habetz

Carolyn Habetz

Samuel Hanchett, Jr.

Dora Ann Hatch

Gert Hawkins

Pat Henry

Celia Hissong

Lucile Horn

Paula Howat

Dorothy Howell

Jane Humble

Nellie Hunt

Andrew Hunter

Peggy & Bill Jenkins

Jane Young Jennings

Charles Johnson

Sandra Karam

Judy Koonce

Simone & Steve Kramer

Barbara Kuhn

Ann & Tommy Laborde

Luke Laborde

Peggy & Lucien Laborde

Alice Lancon

Anne & Gary LeJeune

Karen LeJeune

Kathleen Lemoine

Louisiana Cattlemen's Association

Wilda Lukenbill

Chef Joe Major

Frances Matlock

Julia Mayet

Jan McDade

Tom Merrill

Shannon Millet

Jerry Mitchell

Donna Montgomery

Martha Morrison

Linda Mumphrey

Ellen Murphy

National Cattlemen's Beef Association

David Neal

Tina Ann Nelson

Penny Nichols

Millie & Bob Odom

Nancy Tipton O'Neal

Cary Owen

Ruth Patrick

Alice Peterson

Neila & Cecil Phillips

Nelson Philpot

D'Ann Priakos

Chef Paul Prudhomme

Bill Richardson

Brittany Richardson

Dianne Richardson

Maxine Richardson

Nora & Brandon Richardson

Curt Riché

Raneta Riddle

Grace Rigell

Arlette Rodrigue

Marsha Rogers

Jennifer Rozas

Pamela Rupert

Nick Saban

Janice Saladin

Lisa Savell

A. Kay Singleton

Myrl Sistrunk

Alison Smith

Bill Smith

Bobbie Smith

Majorie Smith

Myrtle Smith

Sherry Smith

Idell Snowden

Dean Soileau

Lola & Bob Soileau

Bryan Songy

Kay Suggs

Alice Teddlie

Maude Thevenot

Barbara Tipton

Judy Turcotte

Kay Vidrine

Patty Vidrine

Georgia Walker

Don Weston

Whitney White

Janine Laborde Williams

Dorothy Wilson

Justin Wilson

Linda Wilson

Margaret Younathan

Dorothy & Ray Young

Leslie & Jesse Young

Denise Zeringue

CORPORATE SPONSORS

Mr. and Mrs. Claud "Buddy" Leach

Mrs. Paula Garvey Manship

Ms. Emogene Pliner

COOKBOOK COMMITTEE

Ganelle Bullock	Gayle Gautreau	Anne LeJeune
Sheryl Carnegie	Carolyn Habetz	Ellen Murphy
	Dorothy Howell	

SPECIAL THANKS

Betty Bagent	David W. Floyd	Chef Joe Major
Bergeron Pecans	Chef John Folse	Dianne Richardson
Mike Cannon	Pat Hegwood	Ruth's Chris Steakhouse

Linda Lightfoot and The Advocate

Louisiana Cattlemen's Association

Louisiana Seafood Marketing Promotion Board

To family and friends of the LSU Agricultural Center, the Cookbook Committee
would like to extend a special thanks for all the recipes submitted, ideas and
props shared, and unfailing enthusiasm for this special project.

Serving Louisiana
VIPs

VIPs

Creole Turtle Soup - 13

Swiss Steak Royale - 14

Harry D. Wilson's Hash - 15

Oven-Baked Roast - 16

Meat, Noodle and Cheese Casserole - 16

Bill's Bobotie - 17

Pork Chops à la Vivi - 18

Chicken Jambalaya - 19

Ro-Tel Chicken - 20

Chicken Enchilada Casserole - 20

21 - Delicious Barbecued Shrimp

22 - Blackened Fish Fillets

23 - The Howard

24 - Nick's Baked Beans

24 - Green Bean Casserole

25 - Sweet Potato Bake

26 - Bread Pudding with Rum Sauce

27 - Rum Sauce

28 - Phillips Phamily Phavorite

Creole Turtle Soup

3 pounds cleaned snapping turtle	1 1/2 cups flour	1/2 teaspoon allspice
2 gallons beef stock	4 cups chopped onions	1/8 teaspoon each nutmeg,
2 onions, cut into quarters	2 cups chopped celery	mace and ground cloves
1 rib celery, chopped	1 cup chopped bell pepper	1/4 cup chow chow
4 garlic cloves	1/4 cup minced garlic	1 tablespoon lemon juice
1 bay leaf	1/2 cup tomato sauce	Salt and cracked pepper
1 cup vegetable oil	2 bay leaves	1/2 cup madeira
	1/2 teaspoon thyme	3 hard-cooked eggs, grated

Mix the turtle, beef stock, 2 onions, 1 rib celery, 4 garlic cloves and 1 bay leaf in a 3-gallon stockpot. Bring to a rolling boil; reduce the heat. Simmer for 1 to 1 1/2 hours or until the turtle is very tender. Strain, reserving 1 gallon of the stock. Discard the vegetables, garlic and bay leaf and coarsely chop the turtle meat.

Heat the oil in a 2-gallon stockpot over medium-high heat. Whisk in the flour until blended. Cook until the roux is golden brown, whisking constantly. Discard the roux if black specks appear and start again. Add 4 cups chopped onions, 2 cups chopped celery, bell pepper and 1/4 cup garlic and mix well. Sauté for 3 to 5 minutes or until the vegetables are tender. Stir in the tomato sauce. Add the reserved stock 1 ladle at a time, mixing well after each addition. Cook until the mixture is of a soup consistency, stirring constantly. The addition of more stock may be required during the cooking process to maintain the desired consistency.

Stir the turtle and next 6 ingredients into the soup mixture. Add the chow chow and lemon juice and mix well. Bring to a low boil; reduce the heat. Simmer for 1 to 1 1/2 hours or until the soup has developed its great flavor. Discard the bay leaves. Season with salt and pepper. Ladle the soup into soup bowls. Drizzle each serving with 1 tablespoon of the wine and sprinkle with 1 tablespoon of the grated egg.

Serves 12

Chef John Folse, CEC, AAC
Chef, Author and Authority on Cajun and Creole cuisine and culture
www.jfolse.com

This Creole specialty is indeed a delicacy in the city of New Orleans. Though created in the bayous by the Cajuns, the original recipe has been changed and adapted to the Creole flavors and may be found in most restaurants in Louisiana.

Swiss Steak Royale

1 1/2 cups flour
4 pounds thick-cut
round steak
4 cups sliced onions
4 cups coarsely chopped celery
1 large green bell pepper,
julienned
1/4 cup vegetable oil

2 cups beef stock or water
1 cup chili sauce
1 (6-ounce) can sliced
mushrooms, drained
2 tablespoons dry mustard
4 teaspoons salt
1/2 teaspoon pepper

Pound the flour into the steak on a hard surface. Sauté the onions, celery and bell pepper in the oil in a heavy ovenproof skillet or Dutch oven. Remove the onion mixture to a bowl using a slotted spoon, reserving the pan drippings.

Brown the steak on both sides in the reserved pan drippings. Stir in the desired amount of stock, chili sauce, mushrooms, dry mustard, salt and pepper. Bake, covered, at 350 degrees for 2 hours. Add the onion mixture and mix well. Bake for 1 hour longer.

Serves 8

Beverly G. Bruce
State Representative Louisiana District #7
Retired Home Economist, LSU AgCenter

Harry D. Wilson's Hash

¹/2 cup vegetable oil
1 cup flour
1 cup chopped onion
¹/2 cup chopped green onions
¹/4 cup chopped fresh parsley
6 cups beef stock or water
2 teaspoons chopped garlic

3 medium potatoes, peeled,
   coarsely chopped
8 to 16 ounces leftover cooked
   beef roast
Salt to taste
Louisiana hot sauce or
   cayenne pepper to taste

Heat the oil in a large heavy high-walled skillet over medium heat. Whisk in the flour. Cook until a dark roux forms, stirring constantly. Stir in the onion, green onions and parsley.

Cook until the onions are tender, stirring frequently. Add 1 cup of the stock and mix until the mixture is the consistency of a thick paste. Stir in the garlic. Add the remaining 5 cups stock and stir until well mixed.

Cut the beef into 1-inch pieces. Add the beef, potatoes, salt and hot sauce and mix well.

Simmer over low heat until the potatoes are tender, stirring occasionally. Serve over hot cooked rice.

Serves 4

Justin Wilson
Louisiana Author, Humorist and TV Host

Permission was granted by Justin Wilson to print this recipe before his death on September 5, 2001.

Oven-Baked Roast

1 (3- to 4-pound) lean chuck roast	2 (10-ounce) cans cream of mushroom soup
Salt and pepper to taste	1 (10-ounce) can French onion soup
Cajun seasoned salt to taste	

Rinse the roast and pat dry. Sprinkle the surface of the roast with salt, pepper and Cajun seasoned salt. Place the roast in a Dutch oven. Pour the soups over the roast; do not add water. Bake, covered, at 350 to 375 degrees for 1^1/4 hours. Turn off the oven. Let stand in the oven with the door closed for 5 hours. Serve over hot cooked rice. The closed oven serves as a big slow cooker, only better!

Serves 8

H. Rouse Caffey, Chancellor Emeritus
LSU AgCenter

Meat, Noodle and Cheese Casserole

12 ounces elbow noodles	1 (16-ounce) can whole kernel corn, drained
1 pound ground beef or turkey	1 pound Cheddar cheese, shredded
1 (8-ounce) jar salsa	1 (8-ounce) package corn chips, crushed
1 tablespoon mayonnaise	

Cook the noodles using package directions; drain. Spray a skillet with nonstick cooking spray. Sauté the ground beef for 7 minutes; drain. Stir in the salsa and mayonnaise. Add the noodles and corn and mix gently. Spoon the ground beef mixture into a 1^1/2-quart baking dish. Sprinkle with the cheese and corn chips. Bake at 350 degrees for 30 minutes.

Serves 6

Bob Odom
Louisiana Commissioner of Agriculture and Forestry

Bill's Bobotie

4 slices white bread
2 tablespoons ginger
2 tablespoons brown sugar
1 tablespoon curry powder
1 tablespoon turmeric
2 teaspoons salt
1/2 teaspoon pepper
1/4 cup (1/2 stick) (or less)
  butter or margarine
5 medium onions, chopped
21/2 pounds ground beef

31/2 ounces raisins
1/4 cup chutney
2 tablespoons apricot jam
2 tablespoons vinegar
2 tablespoons Worcestershire
  sauce
2 tablespoons tomato paste
11/2 cups milk
2 eggs
Lemon leaves (optional)

Soak the bread in enough water to cover in a bowl. Combine the ginger, brown sugar, curry powder, turmeric, salt and pepper in a heavy saucepan and mix well. Heat, stirring frequently to prevent the mixture from burning. Add the butter. Cook until the butter melts, stirring constantly. Sauté the onions in the mixture until tender.

Squeeze most of the moisture from the bread. Add the bread, ground beef, raisins, chutney, jam, vinegar, Worcestershire sauce and tomato paste to the onion mixture and mix well. Cook over medium heat for 20 minutes, stirring frequently. Spoon the ground beef mixture into a greased baking dish.

Whisk the milk and eggs together in a bowl until blended. Pour over the ground beef mixture. Shape each lemon leaf into a funnel shape and pat into the top of the ground beef mixture. Bake at 350 degrees for 45 minutes. Serve immediately.

Serves 8

William Jenkins, President
Louisiana State University System

Pork Chops à la Vivi

4 (6-ounce) lean pork chops, 1 inch thick	4 thin lemon slices
Salt to taste	1/4 cup packed brown sugar
4 thin onion slices	1/4 cup ketchup

Sprinkle both sides of the pork chops with salt. Place in a 9×13-inch baking pan. Arrange 1 onion slice and 1 lemon slice on each pork chop. Top each with 1 tablespoon of the brown sugar and 1 tablespoon of the ketchup.

Bake, covered, at 350 degrees for 1 hour; remove the cover. Bake for 30 minutes longer, basting occasionally with the pan juices.

Serves 4

Ronnie Anderson, President
Louisiana Farm Bureau Federation, Inc.

If you are one of those folks who stays away from pork because you think it is fattening, then rethink this position. Pork is very lean. In addition, pork contains high quality protein, B vitamins, iron, and zinc. Pork is the richest dietary source of thiamine. Shop for pork with a high proportion of lean to fat and bone.

Chicken Jambalaya

1 (4- to 5-pound) hen, cut up	1 teaspoon garlic powder	1 teaspoon garlic powder
1 teaspoon black pepper	8 cups water	1/2 teaspoon MSG
1/2 cup vegetable oil	2 teaspoons salt	(optional)
3 medium onions, chopped	1 teaspoon black pepper	1/4 teaspoon red pepper
	1 teaspoon Tabasco sauce, or 1 tablespoon Cajun Chef	4 cups long grain rice

Sprinkle the hen with 1 teaspoon black pepper. Heat the oil in a 5-quart cast-iron Dutch oven over high heat. Brown the hen on all sides in the hot oil. Reduce the heat to medium when the oil clears. Continue to brown until the bottom of the Dutch oven has a layer of browned bits. Stir in the onions and 1 teaspoon garlic powder.

Cook over low heat until the onions are tender, stirring constantly. Stir in the water. Bring to a boil over high heat. Remove from heat. Let stand for 5 to 10 minutes to allow the oil to rise.

Skim off as much oil as possible. Stir in the salt, 1 teaspoon black pepper, Tabasco sauce, 1 teaspoon garlic powder, MSG and red pepper. Bring to a boil over high heat. Stir in the rice; reduce the heat to low.

Cook, covered, for 10 minutes and stir. Cook, covered, for 20 minutes longer. Serve with a green salad, Cajun white beans and crusty French bread. Substitute 3 pounds cubed boneless pork and 1 pound sliced tube sausage for the hen for Pork and Sausage Jambalaya.

Serves 8

Byron P. Gautreau
World Champion Jambalaya Cook, Gonzales, Louisiana

*Jambalaya is a favorite Louisiana meal. Serve with hot crusty bread
and a salad. It is often prepared outdoors in large quantities in
black pots with game and chicken added. Gonzales, Louisiana, is proud
to be the "Jambalaya Capital of the World."*

Ro-Tel Chicken

6 chicken breasts
12 ounces vermicelli
1 pound Velveeta cheese, cubed
1 (16-ounce) can baby peas, drained
1 (11-ounce) can Ro-Tel tomatoes
1 (2-ounce) jar diced pimentos, drained
1 (4-ounce) jar sliced mushrooms, drained
Salt and pepper to taste

Combine the chicken with enough water to cover in a saucepan. Bring to a boil. Boil until the chicken is tender. Drain, reserving 3 cups of the broth. Chop the chicken, discarding the skin and bones.

Cook the pasta using package directions; drain. Combine the reserved broth, pasta, chicken, cheese, peas, undrained tomatoes, pimentos, mushrooms, salt and pepper in a saucepan and mix well. Cook over low heat until the cheese melts and the liquid is absorbed, stirring frequently. Season with salt and pepper.

Serves 8

Mark Emmert, Chancellor
LSU A&M

Chicken Enchilada Casserole

2 cups shredded baked chicken or turkey
1 (10-ounce) can cream of mushroom soup
1 (10-ounce) can cream of chicken soup
1 cup sour cream
1 large onion, chopped
2 (4-ounce) cans chopped green chiles
12 corn tortillas
1/4 to 1/3 cup vegetable oil
12 ounces Cheddar cheese, shredded

Combine the chicken, soups, sour cream, onion and green chiles in a saucepan and mix well. Cook until heated through, stirring frequently. Fry the tortillas in the oil in a skillet for 1 minute or just until softened. Drain on paper towels.

Layer the tortillas, soup mixture and cheese 1/2 at a time in a buttered 3-quart baking dish. Bake at 350 degrees for 30 minutes.

Serves 8

Pat Henry, Coach
LSU Track and Field

Delicious Barbecued Shrimp

1 1/2 pounds unpeeled shrimp, heads removed
1 cup (2 sticks) margarine
2 tablespoons Worcestershire sauce

1 teaspoon pepper
1 teaspoon parsley flakes
1/2 teaspoon garlic salt
1/2 teaspoon lemon pepper

Arrange the shrimp in a 9×13-inch baking dish. Melt the margarine in a saucepan. Stir in the Worcestershire sauce, pepper, parsley flakes, garlic salt and lemon pepper. Pour the margarine mixture over the shrimp, turning to coat.

Bake at 350 degrees for 25 to 30 minutes or until the shrimp turn pink. Let stand for 10 minutes. Serve unpeeled with warm French bread.

Serves 6

Bill Richardson
Chancellor, LSU AgCenter

Blackened Fish Fillets

| ¼ cup (½ stick) unsalted butter, melted | 6 (8- to 10-ounce) redfish fillets, ½ inch thick, at room temperature | 3 tablespoons Chef Paul Prudhomme's® Blackened Redfish Magic® or Seafood Magic® |

Heat a large cast-iron skillet over high heat until extremely hot. Spread a small amount of the butter on 1 side of 1 of the fillets. Sprinkle with some of the Blackened Redfish Magic®. Arrange the fillet seasoned side down in the hot skillet. Spread a small amount of the butter on the remaining side of the fillet and sprinkle with some of the Blackened Redfish Magic®.

Cook for 4 minutes or until the fillet begins to flake, turning frequently. Repeat the process with the remaining fillets, remaining butter and remaining Blackened Redfish Magic®. Serve hot.

Because this method is simple, any variation will make a dramatic difference. Be sure the skillet is hot enough and absolutely dry. Be sure not to overseason...the herbs and spices should highlight the taste rather than hide it. And you don't want to overcook the fillet...there's a big difference between blackened and burned. Avoid a burned, bitter taste by wiping out the skillet between batches.

Serves 6

Chef Paul Prudhomme
Chef, Author and Louisiana Legend
www.chefpaul.com

If you don't have a commercial hood vent over your stove, this dish will set off every smoke alarm in your neighborhood! It's better to cook it outdoors on a gas grill or a butane burner. Or you can use a charcoal grill, but you'll need to make the coals hotter by giving them extra air. (A normal charcoal fire doesn't get hot enough to "blacken" the fish properly.) Meanwhile, heat your cast-iron skillet for at least ten minutes or as hot as possible on your kitchen stove. When the coals are glowing, use very thick potholders to carefully transfer the hot skillet to the grill.

The Howard

Howard Sauce

1¹/4 cups olive oil
²/3 cup fresh lemon juice
¹/4 cup Worcestershire sauce
2¹/2 tablespoons minced
garlic

2¹/2 tablespoons Mike
Anderson's South Louisiana
Seasoning or seasoned salt

1¹/2 tablespoons crushed
red pepper
1¹/2 tablespoons salt
1¹/2 teaspoons parsley flakes

Baked Fish

8 (16-ounce) fresh white fish fillets
with skin and scales

¹/2 cup (1 stick) butter or margarine, melted
Salt and pepper to taste

For the sauce, combine the olive oil, lemon juice, Worcestershire sauce, garlic, South Louisiana Seasoning, red pepper, salt and parsley flakes in a bowl and mix well.

For the fish, arrange the fillets in a single layer in glass dishes. Reserve 8 tablespoons of the sauce. Pour the remaining sauce over the fillets, turning to coat. Marinate, covered, in the refrigerator for 8 to 10 hours, turning occasionally.

Divide the butter equally between 2 baking dishes. Drain the fillets, reserving the marinade. Arrange the fillets skin side down in the prepared baking dishes. Sprinkle with salt and pepper. Spoon 2 tablespoons of the marinade over each fillet. Bake at 450 degrees for 30 minutes. Spoon 1 tablespoon of the reserved sauce (not used for marinade) on the top of each fillet before serving. Mike Anderson's restaurant in Baton Rouge is a favorite gathering place after LSU graduation ceremonies and sporting events.

Serves 8

Mike Anderson
Former LSU Football Player, Chef and Seafood Specialist

The Howard may also be grilled. Grill the fish skin side down and baste with the Howard Sauce. This dish was named for Mike Anderson's father, Dr. Howard W. Anderson. Dr. Anderson was a Dairy Specialist in the LSU AgCenter for many years.

Nick's Baked Beans

1 pound ground chuck
1 onion, chopped
1/4 cup chopped bell pepper
2 (8-ounce) jars baked beans

1/4 cup ketchup
1/4 cup packed brown sugar
Salt and pepper to taste

Brown the ground chuck with the onion and bell pepper in a skillet, stirring until the ground chuck is crumbly; drain. Stir in the baked beans, ketchup, brown sugar, salt and pepper.

Simmer for 20 minutes, stirring occasionally. Spoon into a serving bowl.

Serves 6

Nick Saban, Coach
LSU Football

Green Bean Casserole

2 (14-ounce) cans cut green
  beans, drained
1/4 cup finely chopped onion
2 tablespoons margarine

1 (10-ounce) can cream of
  mushroom soup
1 (6-ounce) roll garlic cheese,
  sliced

Place the green beans in a 2-quart baking dish. Sauté the onion in the margarine in a saucepan until tender. Stir in the soup and cheese.

Cook until the cheese melts, stirring frequently. Pour the cheese mixture over the beans and stir. Bake at 350 degrees for 30 to 40 minutes or until heated through.

Serves 6

Yvette Girouard, Coach
LSU Softball

Sweet Potato Bake

3 cups mashed cooked
sweet potatoes
1 cup sugar
$^1/_2$ cup milk
$^1/_3$ cup margarine, softened
2 eggs

1 tablespoon vanilla extract
$^1/_2$ teaspoon salt
1 cup packed brown sugar
1 cup ground pecans
$^1/_3$ cup flour
$^1/_2$ cup (1 stick) margarine

Combine the sweet potatoes, sugar, milk, $^1/_3$ cup margarine, eggs, vanilla and salt in a mixing bowl. Beat until blended, scraping the bowl occasionally. Spoon the sweet potato mixture into a greased baking dish.

Combine the brown sugar, pecans, flour and $^1/_2$ cup margarine in a saucepan. Cook over low heat until the margarine melts, stirring frequently. Spread the brown sugar mixture over the prepared layer. Bake at 350 degrees for 30 minutes. You may prepare 1 day in advance and store, covered, in the refrigerator. Bake just before serving. Substitute two 16-ounce cans sweet potatoes for the cooked fresh sweet potatoes if desired.

Coach Brady prefers the sweet potatoes topped with marshmallows. If this is your choice, stir the brown sugar mixture into the sweet potatoes and top the sweet potato mixture with marshmallows. Bake until heated through and the marshmallows are light brown. Four or five large sweet potatoes equal 3 cups mashed cooked sweet potatoes.

Serves 6

John Brady, Coach
LSU Basketball

Bread Pudding with Rum Sauce

1/4 cup (1/2 stick) unsalted
   butter, softened
1 cup confectioners' sugar
1 (1-pound) loaf dry
   French bread
2 cups sugar
4 eggs
2 egg yolks

1 teaspoon vanilla extract
1/2 teaspoon cinnamon
1/8 teaspoon nutmeg
1 quart milk
1 cup shredded sweetened
   coconut
1/2 cup raisins
Rum Sauce (page 27)

Mix the butter and confectioners' sugar in a bowl until creamy. Coat the sides and bottom of a 9×13-inch baking dish with half the butter mixture. Chill the dish in the refrigerator or place in a cool location.

Remove the ends and bottom crust from the bread loaf. Cut the loaf into 1/2-inch slices. Whisk the sugar, eggs, egg yolks, vanilla, cinnamon and nutmeg in a large bowl until blended. Stir in the milk, coconut and raisins. Push the bread slices gently into the milk mixture. Let stand for 30 minutes, turning the slices occasionally.

Pour the bread mixture into the prepared dish and pat down. Push the raisins into the mixture to prevent burning. Heat the remaining butter mixture in a saucepan until melted, stirring frequently. Drizzle over the prepared layer. Arrange the baking dish on a baking sheet and place on an oven rack in the top third of the oven. Bake at 325 degrees for 45 to 50 minutes or until set. Drizzle each serving with some of the Rum Sauce.

Serves 15

Joe Major
Chef/Owner Joe's Dreyfus Store Restaurant, Livonia, Louisiana

Photograph for this recipe appears on page 169.

Rum Sauce

1 cup confectioners' sugar
1/2 cup heavy cream
1/4 cup (1/2 stick) unsalted butter
1 tablespoon dark rum

Combine the confectioners' sugar, cream and butter in a saucepan. Bring to a boil over medium heat, stirring constantly. Remove from heat. Stir in the rum.

Serves 15

In 1902, Theodore Dreyfus purchased a thriving mercantile business from his cousins and, with the help of his wife, children, and several clerks, built a bustling enterprise in an attractive building, circa 1850s, that included an apothecary and the local post office. In 1925, a devastating fire leveled the original building which was then rebuilt to approximately its present state with some modifications and additions through the years.
In 1989, Joe and Diane Major occupied the store building and chose to retain the DREYFUS STORE name, ambience, and goodwill. They wanted the DREYFUS STORE to continue to symbolize warmth and welcome. At the time the DREYFUS STORE became JOE'S DREYFUS STORE RESTAURANT, it was operated by third and fourth generations of the Dreyfus family and is one of the oldest single-family-owned businesses in the state of Louisiana.

Phillips Phamily Phavorite

1/2 cup (1 stick) butter or
margarine
1 (2-layer) package yellow
cake mix
1/2 cup flaked coconut
1 (20-ounce) can sliced apples,
drained

1/2 cup sugar
1 teaspoon cinnamon
1 cup sour cream
2 egg yolks, or 1 egg

Cut the butter into the cake mix in a bowl until crumbly. Stir in the coconut.
Pat the crumb mixture over the bottom and slightly up the sides of an
ungreased 9×13-inch baking pan. Bake at 350 degrees for 10 minutes.

Arrange the apple slices over the warm layer. Sprinkle with a mixture of
the sugar and cinnamon. Whisk the sour cream and egg yolks in a bowl until
blended. Drizzle over the prepared layers. Bake for 25 minutes or until the
edges are light brown.

You may substitute 2 1/2 cups sliced peeled baking apples for the canned
apples. For variety, substitute a white cake mix for the yellow cake mix and
two 16-ounce cans pears for the apples or a chocolate cake mix for the yellow
cake mix and one 29-ounce can peaches for the apples.

Serves 15

Cecil Phillips
President and CEO
LSU Foundation

Serving Louisiana
Appetizers
&
Beverages

Appetizers & Beverages

Avocado and Shrimp Mash

1 small avocado
1 teaspoon minced onion
1 teaspoon lemon juice
1/8 teaspoon salt

1/8 teaspoon pepper
15 butter crackers
15 deveined peeled cooked
shrimp

Mash the avocado in a bowl. Stir in the onion, lemon juice, salt and pepper. Spread the avocado mixture on the crackers. Top each cracker with 1 shrimp. Arrange the crackers on a serving platter.

Serves 15

Crab Canapés

1 cup (2 sticks) margarine,
softened
2 (4-ounce) jars processed
cheese spread
1 tablespoon (heaping)
mayonnaise

1 teaspoon salt
1/4 teaspoon garlic salt
1 pound crab meat, drained
6 English muffins, split, cut
into quarters

Combine the margarine, cheese spread, mayonnaise, salt and garlic salt in a bowl and mix well. Fold in the crab meat.

Spread the crab meat mixture on the cut side of each muffin quarter. Arrange on a baking sheet. Bake at 350 degrees for 10 minutes. Serve immediately.

Makes 4 dozen canapés

Chunky Avocado Dip

5 avocados
3/4 medium onion, chopped
1 tomato, peeled, chopped
1/4 cup fresh lemon juice
1 to 1 1/2 teaspoons garlic salt
1/2 teaspoon pepper, or to taste

7 tablespoons (heaping)
mayonnaise
1 teaspoon Worcestershire
sauce
1 teaspoon Tabasco sauce, or
to taste

Chop the avocados and place in a bowl; do not mash. Add the onion, tomato, lemon juice, garlic salt and pepper and mix gently. Fold in the mayonnaise, Worcestershire sauce and Tabasco sauce.

Chill, covered, for 2 hours or longer. Serve as an appetizer with corn chips or as a salad. Add chopped steamed shrimp for variety.

Serves 10

Don't cut an avocado until it is ripe. Once cut, coat the surface with lemon or lime juice to prevent discoloration. To test for ripeness, squeeze gently…a ripe avocado will yield to gentle pressure. Avocados contain vitamins A, C, and E, folic acid, niacin, thiamine, and riboflavin, as well as being one of the largest suppliers of potassium in the diet. Avocados contain no cholesterol, but are one of the few fruits that contain fat.

Best Crab Dip

2 large onions, finely chopped
6 tablespoons margarine
8 ounces cream cheese, softened
1 pound fresh crab meat, drained
1 tablespoon Worcestershire sauce
1 tablespoon finely chopped parsley
1/4 teaspoon garlic powder
1/4 teaspoon seasoned salt

Sauté the onions in the margarine in a skillet until caramelized. Add the cream cheese. Cook until cream cheese melts, stirring constantly. Add the crab meat and stir gently to mix.

Cook just until heated through, stirring occasionally. Remove from heat. Stir in the Worcestershire sauce, parsley, garlic powder and seasoned salt. Serve with assorted party crackers. This is a great way to use leftover crabs after a crab boil.

Serves 30

Hot Crab Dip

1 bunch green onions, chopped
1 medium bell pepper, chopped
1 cup chopped celery
1/2 cup (1 stick) butter or margarine
1 (10-ounce) can cream of mushroom soup
2 (8-ounce) cans water chestnuts, drained, sliced
2 (6-ounce) cans crab meat, drained
Salt and pepper to taste

Sauté the green onions, bell pepper and celery in the butter in a large skillet until tender. Stir in the soup, water chestnuts, crab meat, salt and pepper.

Cook over medium heat for 20 minutes, stirring frequently. Spoon the crab dip into a chafing dish. Serve with tortilla chips, party breads and/or assorted party crackers.

Serves 20

Crawfish Dip

1 cup chopped green onions
1 medium onion, chopped
2 garlic cloves, minced
1/2 cup (1 stick) butter or margarine
1 (4-ounce) can mushroom stems and
pieces, drained
2 tablespoons Worcestershire sauce
Salt and pepper to taste
Tabasco sauce or Cajun seasoned salt to taste
Lemon juice to taste
8 to 16 ounces peeled crawfish tails, chopped
1 to 2 tablespoons Wondra flour

Sauté the green onions, onion and garlic in the butter in a saucepan until the onions are tender. Stir in the mushrooms, Worcestershire sauce, salt, pepper, Tabasco sauce and lemon juice. Add the crawfish and mix well.

Sauté until the crawfish are cooked through. Stir in the flour. Cook until thickened, stirring constantly. Spoon the dip into a chafing dish. Serve with assorted party crackers and/or waffle chips. May be frozen, covered, for future use. Reheat in the microwave. A touch of cream sherry may be added if desired.

Serves 16

Hamburg Dip

8 ounces cream cheese, softened
1/4 cup milk or cream
3 tablespoons grated or chopped shallot tops
1 garlic clove, crushed
1 tablespoon lemon juice
1/2 teaspoon salt
1/4 teaspoon Worcestershire sauce
1/4 teaspoon Tabasco sauce

Combine the cream cheese, milk, shallot tops, garlic, lemon juice, salt, Worcestershire sauce and Tabasco sauce in a food processor. Process until smooth, adding additional milk if needed for the desired consistency. Taste and adjust seasonings. Spoon into a serving bowl. Serve with fresh vegetables or assorted party crackers. You may substitute 1/2 teaspoon garlic powder for the garlic clove.

Serves 8

Mexican Dip

1 cup mayonnaise
1 cup shredded
Cheddar cheese
1 cup shredded Monterey
Jack cheese
1 (4-ounce) can chopped
black olives, drained

1/8 teaspoon garlic salt
Cumin to taste
1 to 2 cups shredded lettuce
1 large or 2 small avocados,
chopped
1 or 2 tomatoes, chopped

Combine the mayonnaise, Cheddar cheese, Monterey Jack cheese, half the olives, garlic salt and cumin in a bowl and mix well. Spoon the cheese mixture into a baking dish.

Bake at 350 degrees for 15 minutes. Top with the lettuce, remaining olives, avocado and tomatoes. Serve with corn chips.

Serves 8

Mexicorn Dip

2 (12-ounce) cans
Mexicorn, drained
10 ounces Cheddar cheese,
shredded
1 cup mayonnaise
1 cup sour cream

1 (4-ounce) can chopped
green chiles, drained
3 jalapeño chiles, finely
chopped
2 green onions, chopped
1/8 teaspoon sugar

Combine the corn, cheese, mayonnaise, sour cream, green chiles, jalapeño chiles, green onions and sugar in a bowl and mix well. Chill, covered, for several hours. Serve with corn chips.

Serves 25

Strawberry Dip

1 (13-ounce) jar
marshmallow creme

8 ounces strawberry cream
cheese, softened
20 fresh whole strawberries

Combine the marshmallow creme and cream cheese in a bowl and mix well. Chill, covered, in the refrigerator. Serve with the strawberries. You may substitute your favorite fresh fruit for the strawberries.

Serves 10

Monterey Jack Salsa

4 ounces Monterey
Jack cheese
1 tomato, chopped
1 bunch green onions,
chopped

1 (4-ounce) can chopped
green chiles, drained
1 (4-ounce) can chopped
black olives, drained
1/2 cup Italian salad dressing

Shred the cheese. Let stand until room temperature. Combine the cheese, tomato, green onions, green chiles, olives and salad dressing in a bowl and mix gently. Serve with corn chip scoops.

Serves 10

Spinach and Artichoke Dip

2 (10-ounce) packages frozen chopped spinach
1/2 cup chopped onion
1/2 cup (1 stick) butter
1 (15-ounce) can artichoke hearts, drained
8 ounces cream cheese, softened
8 ounces Monterey Jack cheese, shredded
1 cup sour cream
Chopped or minced garlic to taste
Cajun seasoned salt to taste
1/2 cup grated Parmesan cheese

Cook the spinach using package directions; drain. Cool slightly and squeeze any remaining moisture from the spinach. Sauté the onion in the butter in a skillet until tender.

Mash the artichokes in a bowl. Stir in the spinach, sautéed onion, cream cheese, Monterey Jack cheese, sour cream, garlic and Cajun seasoned salt. Spread the spinach mixture in a shallow microwave-safe dish. Sprinkle with the Parmesan cheese.

Microwave for 5 minutes. Serve warm with tortilla chips. You may add the Parmesan cheese to the spinach mixture instead of sprinkling over the top of the dip.

Serves 25

Dried Beef Cheese Ball

3 (2-ounce) jars dried beef, chopped
24 ounces cream cheese, softened
2 small bunches green onions, chopped
2 tablespoons Worcestershire sauce
1 tablespoon MSG

Reserve 1 cup of the dried beef. Combine the remaining dried beef, cream cheese, green onions, Worcestershire sauce and MSG in a bowl and mix well. Shape the cream cheese mixture into a ball. Roll in the reserved dried beef.

Chill, wrapped in plastic wrap, until serving time. Serve with assorted party crackers.

Serves 20

Armadillo Eggs

15 fresh medium
jalapeño chiles
4 ounces Monterey Jack
cheese, slivered
8 ounces pork sausage
8 ounces Monterey Jack
cheese, shredded

1 1/2 cups biscuit mix
2 envelopes Shake'n Bake
for pork
2 eggs, beaten
Picante sauce (optional)

Cut the jalapeño chiles lengthwise into halves. Discard the seeds and veins.
Soak in cold water in a bowl for a milder flavor if desired. Stuff each jalapeño
chile half with the cheese slivers and press the halves together.

Combine the sausage, shredded cheese and biscuit mix in a bowl. The
dough will be stiff so knead with hands to mix. Pat the dough into 1/4-inch-
thick rounds large enough to enclose the jalapeño chiles. Shape 1 round
around each jalapeño chile and roll between hands into the shape of an egg.

Roll each jalapeño chile in Shake'n Bake, dip in beaten eggs and coat
again with Shake'n Bake. Arrange on an ungreased baking sheet. Bake at 325
degrees for 20 to 25 minutes or until light brown, turning several times. Cut
each crosswise into 5 slices. Serve warm or cold with picante sauce.

Makes 75 slices

*Volunteer Master Gardeners throughout Louisiana are trained by
LSU AgCenter specialists to extend their educational services. AgCenter
horticulture agents provide 40 to 50 hours of training for prospective Master
Gardeners every spring and fall at various locales in the state. Master Gardeners
do not pay for the training. Instead, they commit to 40 hours of volunteer service
to their communities for the next year.*

Bourbon Bites

1 cup packed brown sugar
1 cup bourbon
1 cup chili sauce
3 pounds cocktail sausages

Combine the brown sugar, bourbon and chili sauce in a bowl and mix well. Stir in the sausages. Spoon the sausage mixture into a baking dish.

Bake at 350 degrees for 1 hour or until the sausages are cooked through and hot. Cut the sausages into bite-size pieces. Serve hot with wooden picks. You may prepare in advance and reheat in the microwave just before serving.

Serves 40

Picante Chicken

5 or 6 boneless skinless chicken breasts
1/2 bunch green onion tops, chopped
1/2 cup (1 stick) butter
2 to 3 tablespoons Dijon or Creole mustard
2 to 3 tablespoons picante sauce
Lemon pepper to taste

Cut the chicken into 1-inch pieces. Sauté the green onion tops in the butter in an electric skillet or wok until tender. Stir in the chicken; do not let the pieces overlap. Cook until the chicken is white around the edges. Stir in the Dijon mustard, picante sauce and lemon pepper. Turn the chicken.

Cook until the chicken is cooked through, stirring frequently. Remove the chicken to a platter. Serve with wooden picks. You may also serve this dish over rice. Add a mixture of 1 cup water and 2 tablespoons cornstarch to the chicken mixture and cook until thickened.

Serves 15

Turkey Tarts

Tart Shells

8 ounces cream cheese
3/4 cup (1 1/2 sticks) margarine

2 cups flour

Turkey Filling

1 cup chopped cooked turkey
1/3 cup shredded
Velveeta cheese
1/3 cup mayonnaise
1/3 cup thinly sliced celery

2 tablespoons sliced green
onions
2 tablespoons sour cream
1/4 teaspoon salt
1/4 teaspoon pepper

For the shells, cut the cream cheese and margarine into the flour in a bowl until a soft dough forms. Chill, covered with plastic wrap, for 1 hour. Shape the dough into 36 equal portions. Press each portion over the bottom and up the side of a greased miniature muffin cup. Bake at 400 degrees for 8 to 10 minutes or until golden brown.

For the filling, combine the turkey, cheese, mayonnaise, celery, green onions, sour cream, salt and pepper in a bowl and mix well. Spoon about 1 tablespoon of the turkey mixture into each muffin cup. Bake for 3 to 5 minutes or until the cheese melts. Serve warm.

Makes 3 dozen tarts

Salmon Tarts

4 (9-inch) refrigerated pie pastries
1½ cups half-and-half
4 eggs, beaten
4 ounces salmon, chopped
½ cup shredded Monterey Jack cheese
¼ cup minced green onions
½ teaspoon dillweed
¼ teaspoon salt
⅛ teaspoon pepper

Cut each pie pastry into 14 rounds using a 2½-inch round cutter. Pat each pastry round over the bottom and up the side of a greased miniature muffin cup, trimming the excess pastry. Whisk the half-and-half and eggs in a bowl. Stir in the salmon, cheese, green onions, dillweed, salt and pepper.

Spoon 1 tablespoon of the salmon mixture into each pastry-lined muffin cup. Bake at 375 degrees for 25 to 30 minutes or until set. Remove to a wire rack to cool. You may prepare in advance and store in an airtight container in the freezer for up to 2 weeks. To serve, thaw at room temperature. Arrange the tarts on a baking sheet. Reheat, covered, at 375 degrees for 5 to 10 minutes or until hot.

Makes 56 tarts

Crab and Cream Cheese Bake

8 ounces cream cheese, softened
¼ cup chopped green onions
1 (8-count) can crescent rolls
1 cup lump crab meat
1 egg yolk, beaten
½ teaspoon dillweed

Combine the cream cheese and green onions in a bowl and mix well. Unroll the crescent roll dough on a greased baking sheet; do not separate the triangles. Press the dough into an 8×11-inch rectangle.

Spoon the crab meat lengthwise down the center of the rectangle. Spread the cream cheese mixture over the crab meat. Fold the long edges of the rectangle over the cream cheese mixture, slightly overlapping. Pinch the edges to seal.

Arrange seam side down on the baking sheet. Brush the top lightly with the egg yolk and cut slits. Bake at 350 degrees for 20 to 22 minutes or until golden brown and flaky. Cut into twelve slices. Serve warm.

Serves 12

Sausage and Cheese Squares

1 (8-count) can crescent rolls
1 pound bulk pork sausage
2 cups shredded Monterey Jack cheese
3/4 cup milk
4 eggs, lightly beaten
1/4 cup chopped green bell pepper
1/4 teaspoon pepper
1/4 teaspoon oregano

Unroll the crescent roll dough and separate into 2 large rectangles. Press the rectangles over the bottom and 1/2 inch up the sides of an ungreased 9×13-inch baking pan, pressing the edges and perforations to seal.

Brown the sausage in a skillet, stirring until crumbly; drain. Spoon the sausage over the prepared layer. Sprinkle with the cheese. Whisk the milk, eggs, bell pepper, pepper and oregano in a bowl until mixed. Pour the egg mixture over the prepared layers.

Bake at 425 degrees for 20 to 25 minutes or until set. Cut into bite-size squares. Cut into larger squares and serve with fruit for breakfast or with a salad for lunch or dinner. Very versatile dish.

Makes 4 dozen squares

Sugared Pecans

1 pound pecans, shelled
2 cups sugar
3/4 cup water

Combine the pecans, sugar and water in a skillet and mix well. Cook until the liquid evaporates, the pecans are coated and the sugar has begun to caramelize, stirring constantly with a wooden spoon. If you cook too long the sugar will turn to brown syrup.

Pour the pecans onto a sheet of foil and separate with a wooden spoon or fork. Let stand until cool.

Freshly shelled pecans should be allowed to dry before storing. Allow the pecans to stand at room temperature in open pans for at least one week before storing. This drying process also improves the flavor. If not dry, the pecans are more likely to mold, even in your refrigerator. Always store in airtight containers to prevent the absorption of odors from other foods.

Serves 8

Bourbon Slush

2 tea bags
1 cup boiling water
3/4 cup sugar
3 1/2 cups water
1/2 cup bourbon
1 (6-ounce) can frozen orange juice
concentrate
1/2 (6-ounce) can frozen lemonade concentrate

Steep the tea bags in the boiling water in a 4-cup heatproof measuring cup for 3 to 5 minutes; discard the tea bags. Add the sugar, stirring until dissolved. Combine the tea mixture, water, bourbon, orange juice concentrate and lemonade concentrate in a freezer container and mix well.

Freeze for at least 48 hours before serving, stirring 2 to 3 times during the first 24 hours; be sure to "drag" the bottom of the container to distribute the alcohol. Remove from freezer 10 to 15 minutes before serving. Spoon into glasses.

Serves 8

Cocktail Slush

1 (46-ounce) can pineapple juice
1 liter lemon-lime soda
1 1/2 cups vodka, or to taste
1 (20-ounce) can pineapple chunks
1 (12-ounce) can frozen orange juice
concentrate
1/2 (8-ounce) jar maraschino cherries
Juice of 1 lemon

Combine the pineapple juice, soda, vodka, undrained pineapple chunks, orange juice concentrate, undrained cherries and lemon juice in a freezer container and mix well. Freeze, covered, for 8 to 10 hours. Let stand at room temperature until of a slush consistency. Spoon into glasses. You may substitute rum for the vodka.

Serves 25

Mimosa

2 cups orange juice, chilled
2 cups cranberry juice
cocktail, chilled

1 (750-milliliter) bottle
sparkling wine, chilled

Mix the orange juice and cranberry juice in a pitcher; place in the freezer if desired to get icy cold. Fill stemmed glasses half full with the wine. Pour enough of the fruit juice mixture into each glass to fill; stir.

Serves 10

Photograph for this recipe appears on page 11.

Tiger Punch

2 (6-ounce) cans frozen
lemonade concentrate
1 (46-ounce) can
pineapple juice

1 tablespoon almond extract
Fresh mint leaves (optional)
Lemon slices (optional)

Prepare the lemonade concentrate using can directions. Pour the lemonade into a punch bowl. Stir in the pineapple juice and almond extract. Garnish with mint leaves and lemon slices. Ladle into punch cups.

Serves 25

Golden Punch

1/2 cup sugar
8 to 10 cups water
2 (6-ounce) cans frozen orange
   juice concentrate
2 (6-ounce) cans frozen
   lemonade concentrate

1/2 (46-ounce) can pineapple
   juice
Ice cubes or ice ring
1 1/2 to 2 quarts ginger ale,
   chilled

Dissolve the sugar in 1 cup of the water in a large container. Add the orange juice concentrate, lemonade concentrate and pineapple juice. Rinse the frozen juice cans with part of the water. Stir in the remaining water.

Pour the punch mixture into a punch bowl. Add ice cubes or an ice ring. Stir in the ginger ale. Ladle into punch cups.

Serves 50

Rosé Sparkle

3 (10-ounce) packages frozen
   raspberries
3 (1/2-gallon) bottles rosé,
   chilled
1 (2-liter) bottle diet lemon-
   lime soda, chilled

2 (6-ounce) cans frozen pink
   lemonade concentrate
Fresh mint leaves

Combine the raspberries, rosé, soda and lemonade concentrate in a punch bowl and mix gently. Garnish with mint leaves. Ladle into punch cups.

Serves 50

Orange Breakfast Shake

1 cup 2% milk
1 ripe banana, sliced
1/4 cup frozen orange juice
concentrate

2 teaspoons sugar, or
1 packet artificial sweetener

Combine the milk, banana, orange juice concentrate and sugar in a blender. Process until smooth. Pour into glasses.

Serves 2

Hot Spiced Tea

2 cups orange instant
breakfast drink mix
2 cups sugar
1 large envelope sweetened
lemonade mix

1/2 cup instant tea granules
2 teaspoons cinnamon
1 1/2 teaspoons nutmeg
1 1/2 teaspoons ground cloves

Sift the drink mix, sugar, lemonade mix, tea granules, cinnamon, nutmeg and cloves into a large bowl and mix well. Store the hot tea mix in an airtight container. To serve, mix 3 to 4 teaspoons of the mixture with 1 cup hot water in a mug.

Serves 25

Serving Louisiana
Soups
&
Salads

Soups & Salads

Chili - 49	57 - Shrimp and Okra Gumbo
Mémère JuJu's Broccoli Soup - 50	58 - Cranberry Salad
Cabbage and Sausage Soup - 51	58 - Marinated Fruit Salad
Cheese Soup - 51	59 - Apricot Salad
Chicken Noodle Soup - 52	59 - Southern Pretzel Strawberry Salad
Corn and Crab Soup - 52	60 - Nine-Day Coleslaw
Cream of Crawfish Soup - 53	60 - Herbed Tomato Salad
French Potato Soup - 53	61 - Bleu Cheese Chicken Salad
Tortilla Soup - 54	62 - Polynesian Chicken Salad
Vegetable Beef Soup - 54	63 - Mexican Fiesta Chicken and Rice Salad
Cajun Chicken and Sausage Gumbo - 55	64 - Seafood Rice Salad
Seafood Filé Gumbo - 56	64 - Shrimp and Pasta Salad

Chili

1 pound ground beef or turkey
1 onion, chopped
1 bell pepper, chopped
3 garlic cloves, minced
1 (28-ounce) can whole
tomatoes
1 (16-ounce) can red
beans, drained
1 (8-ounce) can tomato sauce
2 to 4 tablespoons chili
powder, or to taste

2 teaspoons salt
1 to 2 teaspoons sugar
1 teaspoon oregano
1 teaspoon cumin
1 teaspoon marjoram
1 teaspoon cinnamon
1 teaspoon paprika
1/2 teaspoon red pepper
1 bay leaf

Brown the ground beef with the onion, bell pepper and garlic in a large saucepan, stirring until the ground beef is crumbly; drain. Stir in the undrained tomatoes, beans, tomato sauce, chili powder, salt, sugar, oregano, cumin, marjoram, cinnamon, paprika, red pepper and bay leaf.

Simmer for 1 hour or until of the desired consistency, stirring occasionally. Discard the bay leaf. Ladle into chili bowls.

Serves 6

Mémère JuJu's Broccoli Soup

3 boneless skinless
chicken breasts
2 onions, chopped
1/2 cup (1 stick) margarine
1 (14-ounce) can
chicken broth
1 (10-ounce) can cream of
chicken soup
1 (10-ounce) can cream of
broccoli soup
1 (10-ounce) can cream of
celery soup

1 (10-ounce) can cream of
mushroom soup with
roasted garlic
4 soup cans milk
1 1/2 pounds fresh broccoli,
trimmed, steamed, drained
1 pound mild or hot Mexican
Velveeta cheese, cubed
1 1/2 cups sliced fresh
mushrooms

Spray the chicken with nonstick cooking spray. Arrange in a baking pan. Bake at 350 degrees for 1 hour. Cool slightly and coarsely chop.

Sauté the onions in the margarine in a saucepan until caramelized. Stir in the broth, soups and milk. Bring to a boil, stirring constantly. Add the broccoli, cheese and mushrooms and mix gently; reduce the heat.

Simmer for 40 minutes, stirring occasionally. Stir in the chicken. Simmer for 20 minutes longer, stirring occasionally. Ladle into soup bowls. You may substitute two 10-ounce packages frozen chopped broccoli for the fresh broccoli.

Serves 12

Cabbage and Sausage Soup

1 pound smoked pork sausage with garlic
2 cups chopped onions
6 garlic cloves, chopped
6 tablespoons vegetable oil
5 (14-ounce) cans beef broth
1 (15-ounce) can kidney beans
1 (20-ounce) bottle ketchup
1 head green cabbage, chopped
12 small new potatoes, cut into bite-size pieces
1/4 cup vinegar

Cut the sausage into bite-size pieces. Sauté the onions and garlic in the oil in a skillet. Add the sausage. Sauté until the sausage is light brown; drain. Spoon the sausage mixture into a stockpot. Stir in the broth, undrained beans, ketchup, cabbage, potatoes and vinegar. Bring to a boil, stirring frequently; reduce the heat.

Simmer for 1 hour, stirring occasionally. Ladle into soup bowls. You may store, covered, in the refrigerator for up to 1 week.

Serves 8

Cheese Soup

1 quart water
4 chicken bouillon cubes
1 cup chopped onion
1 cup chopped celery
2 (10-ounce) packages frozen chopped broccoli
1 pound Velveeta cheese, cubed
2 (10-ounce) cans cream of chicken soup

Combine the water and bouillon cubes in a heavy saucepan. Add the onion and celery. Bring to a boil. Boil for 10 minutes, stirring occasionally. Stir in the broccoli; reduce the heat.

Cook for 10 minutes or until the broccoli is tender, stirring occasionally. Add the cheese and soup and mix well. Cook until the cheese melts, stirring constantly. Ladle into soup bowls. You may substitute any vegetable for the broccoli.

Serves 6

Chicken Noodle Soup

1 (3 1/2- to 4-pound) chicken, cut up
10 to 12 cups water
2 cups chopped onions
2 cups chopped celery
2 teaspoons seasoned salt
1 (16-ounce) package frozen mixed vegetables
1 (16-ounce) package frozen corn
1 envelope vegetable soup mix
16 ounces thin spaghetti

Combine the chicken, water, onions, celery and seasoned salt in a stockpot. Bring to a boil; reduce the heat. Simmer, covered, for 1 1/2 hours or until the chicken is tender. Remove the chicken to a platter, reserving the broth. Chop the chicken, discarding the skin and bones.

Add the mixed vegetables, corn and soup mix to the reserved broth. Simmer, covered, for 10 minutes. Stir in the chicken and pasta. Cook over low heat for 15 minutes, stirring occasionally. Adjust the seasonings. Ladle into soup bowls.

Serves 20

Corn and Crab Soup

1 large onion, chopped
4 (10-ounce) cans reduced-fat
cream of potato soup
1 (10-ounce) can reduced-fat
cream of celery soup
2 cups drained cooked whole kernel corn
8 ounces reduced-fat cream cheese, cubed
2 (6-ounce) cans crab meat, drained
1 cup reduced-fat sour cream
1 tablespoon butter sprinkles
Salt and pepper to taste
Chopped fresh parsley to taste

Sauté the onion in a heavy 4- to 6-quart saucepan sprayed with nonstick cooking spray until the onion is tender. Add the soups and mix well. Cook over low heat for 15 minutes, stirring occasionally. Stir in the corn, cream cheese and crab meat.

Cook over low heat until the cream cheese melts, stirring occasionally. Stir in the sour cream, butter sprinkles, salt, pepper and parsley. Simmer for 15 minutes longer, stirring occasionally. Ladle into soup bowls.

Serves 10

Cream of Crawfish Soup

1 pound peeled crawfish tails
1/2 bunch green onions, coarsely chopped
1/2 cup grated onion
1/2 cup (1 stick) butter
1/2 cup flour
2 cups chicken stock, heated
2 cups half-and-half
2 cups heavy cream
2 teaspoons red pepper
2 teaspoons garlic powder
2 teaspoons onion powder

Process the crawfish tails and green onions in a food processor until ground. Sauté the onion in the butter in a large heavy saucepan for 5 minutes. Add the flour and mix well. Cook for 2 minutes or until thickened, stirring constantly. Stir in the hot chicken stock.

Simmer for 5 minutes, stirring constantly. Stir in the crawfish mixture. Simmer for 5 minutes, stirring frequently. Add the half-and-half and heavy cream and mix well. Simmer for 5 minutes, stirring occasionally. Stir in the red pepper, garlic powder and onion powder. Ladle into soup bowls.

Serves 6

French Potato Soup

4 or 5 green onion bulbs, chopped, or
1/2 cup sliced white onion
1/4 cup (1/2 stick) margarine
3 cups chicken broth
2 cups thinly sliced potatoes
1 cup milk or half-and-half
Salt and pepper to taste
Minced fresh parsley or chives to taste

Sauté the green onions in the margarine in a saucepan for 5 minutes or until tender. Add the broth and potatoes and mix well. Simmer, covered, for 30 minutes or until the potatoes are tender, stirring occasionally. Cool slightly. Force the potato mixture through a fine sieve or process in a blender until smooth.

Return the potato mixture to the saucepan. Stir in the milk, salt and pepper. Simmer until heated through, stirring frequently. Serve hot or cold, but it is better cold. Sprinkle each serving with parsley or chives. The flavor of the soup is enhanced if prepared 1 day in advance and stored, covered, in the refrigerator.

Serves 4

Tortilla Soup

3 boneless skinless chicken breasts, (optional)
1 onion, chopped
1 green bell pepper, chopped
2 garlic cloves, minced
1 to 2 tablespoons vegetable oil
1 (16-ounce) can whole kernel corn
2 (14-ounce) cans stewed tomatoes
1 (14-ounce) can beef broth
1 1/2 cups water
1 cup sliced zucchini
1 (4-ounce) can green chiles, drained, chopped
1/2 cup picante sauce or salsa
1 teaspoon cumin
6 tortillas, cut into 1/2-inch strips
1/2 cup shredded reduced-fat Cheddar cheese

Cut the chicken into 1-inch pieces. Sauté the onion, bell pepper and garlic in the oil in a 6-quart Dutch oven until tender. Stir in the undrained corn, undrained tomatoes, broth, water, zucchini, green chiles, picante sauce, cumin and chicken. Bring to a boil; reduce the heat.

Simmer, covered, for 1 hour, stirring occasionally. Add the tortillas and mix well. Simmer for 5 minutes longer, stirring occasionally. Ladle into soup bowls. Sprinkle each serving with some of the cheese.

Serves 12

Vegetable Beef Soup

8 ounces lean ground beef or turkey
3 cups water
1 bay leaf (optional)
1 large carrot, sliced
2 ribs celery, chopped
1 small onion, chopped
2 cups mixed vegetables
2 cups canned tomatoes
3/4 cup chopped potato
1/4 cup rice
2 beef bouillon cubes
Lite salt and pepper to taste

Brown the ground beef in a 4-quart saucepan, stirring until crumbly; drain. Add the water and bay leaf and mix well; cover. Bring to a boil. Stir in the carrot, celery, onion, mixed vegetables, tomatoes, potato, rice, bouillon cubes, lite salt and pepper.

Bring to a boil; reduce the heat. Simmer for 20 to 30 minutes or until the vegetables are tender, stirring occasionally. Discard the bay leaf. Ladle into soup bowls.

Serves 6

Cajun Chicken and Sausage Gumbo

4 quarts water
1 pound smoked pork
sausage, sliced
1 cup Basic Traditional Roux
(below)
2 medium onions, chopped
1 bell pepper, chopped
1 tablespoon Cajun
seasoned salt

4 garlic cloves, chopped
6 boneless skinless chicken
thighs, cut into halves
1/2 cup chopped green onions
1/2 cup chopped fresh parsley
Cajun seasoned salt to taste
5 cups hot cooked rice

Bring the water to a boil in a 6-quart stockpot. Add the sausage, roux, onions, bell pepper, 1 tablespoon Cajun seasoned salt and garlic. Bring to a boil; reduce the heat. Simmer for 1 hour, stirring occasionally.

Add the chicken to the gumbo. Simmer for 30 minutes, stirring occasionally. Stir in the green onions and parsley. Simmer for 10 minutes longer, stirring occasionally. Season with Cajun seasoned salt to taste. Ladle the gumbo over the hot cooked rice in soup bowls. Serve with hot sauce.

Serves 10

Roux is the base for all gumbos, gravies and stews. To prepare Basic Traditional Roux, heat 1 cup butter, shortening or bacon drippings in a cast-iron Dutch oven or skillet. Add 1 cup flour gradually, stirring constantly. Cook until dark brown, being careful not to burn. If there is the slightest indication of overbrowning, dispose of the roux and start over. Even a slightly burned roux will ruin a savory dish. Add seasonings and stock to make various sauces and gravies. You may store the roux in the refrigerator or freeze for future use. This recipe yields one cup.

Seafood Filé Gumbo

3 tablespoons (slightly
heaping) shortening
3 tablespoons (slightly
heaping) flour
8 to 10 small okra pods,
chopped
1 bunch green onions, chopped
3 quarts water
3 pounds deveined peeled
shrimp

1 pound fresh crab meat
(optional)
1 pint oysters (optional)
Salt and pepper to taste
2 tablespoons tomato paste
1 to 2 tablespoons filé powder
4 cups hot cooked rice

Heat the shortening in a cast-iron skillet. Add the flour gradually, stirring constantly. Cook over medium to medium-low heat until the roux is dark brown, stirring constantly. Add the okra and green onions and mix well. Sauté for 5 minutes.

Bring the water to a boil in a 6-quart stockpot over high heat. Add the okra mixture and mix well; reduce the heat. Simmer for 15 minutes, stirring occasionally. Stir in the shrimp, crab meat, oysters, salt and pepper. Cook for 15 minutes, stirring occasionally. Stir in the tomato paste. Add the filé powder, stirring gently until the filé powder dissolves. Remove from the heat. Let stand, covered, for 10 to 15 minutes. Ladle over the rice in soup bowls.

Serves 8

Photograph for this recipe appears on page 47.

*Gumbo comes from the African word, kingumbo, which means okra.
Africans first brought okra, a crucial ingredient in gumbo recipes, to Louisiana.
Louisiana's native Americans, the Choctaw Indians, also contributed to the pot
with filé—a powder made from dried, ground sassafras leaves which can
also serve to flavor and thicken the gumbo.*

Shrimp and Okra Gumbo

3 slices bacon
3 tablespoons flour
1 pound fresh tender okra
pods, trimmed, chopped
1 cup chopped onion
1/2 cup chopped green
bell pepper
2 garlic cloves, minced
1 (16-ounce) can tomatoes
2 cups water

1 (8-ounce) can tomato sauce
2 teaspoons salt
1 teaspoon sugar
1/4 teaspoon pepper
1/4 teaspoon Tabasco sauce
1/8 teaspoon MSG
2 (12-ounce) packages frozen
deveined peeled shrimp
3 cups hot cooked rice

Fry the bacon in a cast-iron skillet or Dutch oven until crisp. Remove the bacon to a plate, reserving the bacon drippings. Add the flour gradually to the reserved bacon drippings, stirring constantly. Cook over low heat until dark brown, stirring constantly. Stir in the okra, onion, bell pepper and garlic.

Cook until the vegetables are tender, stirring frequently. Crumble the bacon into the vegetable mixture and mix well. Spoon the mixture into a stockpot. Stir in the undrained tomatoes, water, tomato sauce, salt, sugar, pepper, Tabasco sauce and MSG.

Simmer, covered, for 1 hour or until thickened, stirring occasionally. Stir in the shrimp. Simmer for 30 minutes or until the shrimp turn pink, stirring occasionally. Taste and adjust the seasonings. Ladle over the rice in soup bowls. Serve with hot crusty French bread.

Serves 6

Cranberry Salad

1 cup fresh cranberries	15 miniature marshmallows
1 cup sugar	1 cup chopped celery
1 cup water	1 cup chopped apples
1 (6-ounce) package cherry gelatin	1 cup chopped pecans

Combine the cranberries, sugar and water in a saucepan. Cook until the cranberries pop, stirring occasionally. Stir in the gelatin. Add the marshmallows and stir until the marshmallows melt. Let stand until cool but not set.

Add the celery, apples and pecans to the gelatin mixture and mix well. Spoon into a mold. Chill, covered, until set.

Serves 8

Marinated Fruit Salad

1 (21-ounce) can peach pie filling	1 (10-ounce) package frozen sliced strawberries
1 (15-ounce) can pineapple chunks, drained	3 bananas, sliced
1 (11-ounce) can mandarin oranges, drained	

Combine the pie filling, pineapple chunks, mandarin oranges, strawberries and bananas in a bowl and mix gently. Chill, covered, for 8 to 10 hours.

Serves 10

Apricot Salad

1 (15-ounce) can crushed pineapple
1 (6-ounce) package dried apricots
2/3 cup sugar
1 (6-ounce) package apricot gelatin
1 cup boiling water
8 ounces cream cheese, softened
2 (3-ounce) jars strained apricots (baby food)
8 ounces whipped topping
1/2 cup chopped pecans

Drain the pineapple, reserving the juice. Pour the reserved juice over the dried apricots in a bowl. Let stand until softened. Snip the apricots into small pieces with kitchen shears.

Combine the sugar and gelatin in a heatproof bowl and mix well. Add the boiling water, stirring until dissolved. Stir the warm gelatin mixture into the cream cheese in a bowl. Add the strained apricots and pineapple and mix well. Let stand until cool.

Add the softened dried apricots to the gelatin mixture. Fold in the whipped topping and pecans. Spoon the gelatin mixture into a 9×13-inch dish. Chill, covered, until set.

Serves 15

Southern Pretzel Strawberry Salad

2 cups pretzels, crushed
3/4 cup (1 1/2 sticks) margarine, melted
3 tablespoons sugar
16 ounces whipped topping
8 ounces cream cheese, softened
1 cup sugar
1 (6-ounce) package strawberry gelatin
2 cups boiling water
1 (10-ounce) package frozen strawberries
Fresh strawberries (optional)

Mix the pretzels, margarine and 3 tablespoons sugar in a bowl. Pat the crumb mixture over the bottom of a 9×13-inch baking pan. Bake at 400 degrees for 8 to 10 minutes or until light brown. Let stand until cool.

Beat whipped topping, cream cheese and 1 cup sugar in a mixing bowl until smooth, scraping the bowl occasionally. Spread the cream cheese mixture over the baked layer. Dissolve the gelatin in the boiling water in a heatproof bowl. Stir in the frozen strawberries. Let stand until partially set. Spoon over the prepared layers. Chill, covered, until set. Garnish with fresh strawberries.

Serves 15

Nine-Day Coleslaw

1 cup sugar
1 medium head cabbage, coarsely shredded
1 medium onion, chopped
1 green bell pepper, chopped
4 ribs celery, chopped
1 (4-ounce) jar chopped pimentos, drained
1 cup vegetable oil
1 cup vinegar
2 tablespoons salt
1 teaspoon dry mustard
1 teaspoon celery seeds

Set aside 2 tablespoons of the sugar. Toss the cabbage, onion, bell pepper, celery and pimentos with the remaining sugar in a bowl. Combine 2 tablespoons sugar, oil, vinegar, salt, dry mustard and celery seeds in a saucepan and mix well. Bring to a boil, stirring constantly.

Pour the hot dressing over the cabbage mixture and mix well. Let stand until cool. Store, covered, in the refrigerator for 24 hours. You may serve immediately but the flavor is enhanced the longer the coleslaw is chilled and the slaw is still good after nine days. Do not use a food processor or grater to shred the cabbage.

Serves 12

Herbed Tomato Salad

6 medium to large tomatoes, sliced
2/3 cup vegetable oil
1/4 cup vinegar
1 or 2 garlic cloves, minced
1 teaspoon salt
1/2 teaspoon thyme
1/4 teaspoon pepper
1/4 cup minced fresh parsley
1/4 cup sliced green onions with tops

Arrange the sliced tomatoes in a dish. Whisk the oil, vinegar, garlic, salt, thyme and pepper in a bowl. Drizzle over the tomatoes. Marinate at room temperature for several hours. Sprinkle each serving with parsley and green onions.

Serves 12

Bleu Cheese Chicken Salad

4 (4-ounce) boneless skinless
chicken breasts
2 teaspoons dillweed
1/4 teaspoon salt
1/4 teaspoon white pepper
6 cups torn romaine
1 medium green bell pepper,
cut into rings

1 medium cucumber, sliced
2 tablespoons olive oil
1 tablespoon Dijon mustard
1 tablespoon balsamic vinegar
2 medium tomatoes,
cut into eighths
2 ounces bleu cheese,
crumbled

Spray both sides of the chicken with nonstick cooking spray. Sprinkle evenly with the dillweed, salt and white pepper. Heat a cast-iron skillet over high heat until hot. Sauté the chicken in the hot skillet until brown on both sides. Bake at 350 degrees for 10 minutes. Cool slightly and cut the chicken diagonally into thin slices.

Toss the lettuce, bell pepper and cucumber in a bowl. Whisk the olive oil, Dijon mustard and balsamic vinegar in a bowl. Add to the lettuce mixture and toss to coat.

To serve, divide the lettuce mixture evenly among 8 salad plates. Top each serving with sliced chicken and tomato wedges. Sprinkle with the bleu cheese.

Serves 8

Polynesian Chicken Salad

2 cups chopped cooked
chicken breasts
1 1/2 cups chopped celery
1 1/2 cups drained canned
pineapple chunks
1/2 cup sliced almond, toasted
1/4 cup shredded carrot

1/2 cup mayonnaise
1/4 cup sour cream
1 teaspoon curry powder
1 teaspoon fresh lemon juice
1/2 teaspoon salt
Lettuce leaves
Minced fresh parsley to taste

Combine the chicken, celery, pineapple, almonds and carrot in a bowl and mix well. Mix the mayonnaise, sour cream, curry powder, lemon juice and salt in a bowl. Add the mayonnaise mixture to the chicken mixture and toss gently to coat. Chill, covered, in the refrigerator until serving time.

To serve, spoon the chicken mixture onto lettuce-lined salad plates. Sprinkle with parsley. Serve immediately. You may substitute green grapes for the pineapple chunks. Do not grate the carrot.

Serves 6

Food is a big business in Louisiana. Not only do Louisianians appreciate their cuisine, but so do the out-of-staters who contribute to the multimillion dollar tourist business. But the bubble would soon burst if food safety were to be a problem. It is not, thanks to the LSU AgCenter's professional safety training to food processors of all types. The New Orleans Jazz Festival, a popular musical event, requires all food handlers to be trained in safety by AgCenter specialists.

Mexican Fiesta Chicken and Rice Salad

2$\frac{1}{2}$ cups water
1 cup brown rice
1 teaspoon seasoned salt
1 bell pepper, chopped
1 bunch green onions,
  chopped
2 boneless skinless chicken
  breasts, cooked, sliced
2 (15-ounce) cans black beans,
  rinsed, drained

1 (15-ounce) can Mexicorn,
  drained
1 cup salsa
2 large lettuce leaves
1 cup shredded Pepper Jack
  cheese
$\frac{1}{2}$ cup nonfat sour cream
1 tomato, cut into wedges
Jalapeño chiles (optional)

Combine the water, brown rice and seasoned salt in a microwave-safe dish and mix well. Microwave, covered, on High for 15 minutes. Microwave, covered, on Low for 30 minutes longer. Let stand covered until ready to use.

Sauté the bell pepper and green onions in a skillet sprayed with nonstick cooking spray until tender. Add the chicken and mix well. Stir in the beans, corn and salsa.

Cook until the vegetables are tender, stirring frequently. Add the brown rice and mix well. Cook until heated through, stirring occasionally. Spoon the chicken mixture onto a lettuce-lined serving platter. Sprinkle with the cheese and top with dollops of the sour cream. Arrange the tomato and jalapeño chiles around the salad. Serve with baked tortilla chips.

Serves 8

Seafood Rice Salad

8 ounces frozen peeled cooked shrimp, thawed
3 cups cooked rice, at room temperature
1 cup finely chopped celery
1/2 cup finely chopped onion
1/4 cup finely chopped bell pepper
1/4 cup chopped pimentos
3 hard-cooked eggs, chopped
1 cup mayonnaise
1 tablespoon lemon juice
Salt and pepper to taste
Mixed salad greens to taste
Tomato wedges (optional)

Combine the shrimp, rice, celery, onion, bell pepper, pimentos and eggs in a bowl and mix gently. Stir in a mixture of the mayonnaise and lemon juice. Season with salt and pepper.

Chill, covered, in the refrigerator. Spoon the salad onto a platter lined with mixed salad greens. Garnish with tomato wedges.

Serves 6

Shrimp and Pasta Salad

16 ounces cavatappi (corkscrew pasta)
1/2 cup chopped celery
1/2 cup mayonnaise
2 tablespoons prepared mustard
1 tablespoon milk
1/2 teaspoon salt
1/8 teaspoon sugar
Freshly ground pepper to taste
1 to 2 teaspoons Cajun seasoned salt
1 1/2 pounds shrimp, cooked, peeled, deveined

Cook the pasta using package directions in a saucepan; drain. Stir in the celery. Combine the mayonnaise, prepared mustard, milk, salt, sugar and pepper in a bowl and mix well. Add the mayonnaise mixture to the pasta mixture and mix well. Sprinkle with the Cajun seasoned salt. Fold in the shrimp. Spoon into a serving bowl.

Chill, covered, in the refrigerator. The flavor of the salad is enhanced if prepared 1 day in advance and stored, covered, in the refrigerator.

Serves 8

Meat, Poultry & Game

Rib-Eye Roast with Oven-Browned Vegetables

2 tablespoons minced fresh
rosemary leaves
4 garlic cloves, crushed
1 teaspoon salt
1 teaspoon dry mustard
1 teaspoon cracked
black pepper
1 (4-pound) boneless beef
rib-eye roast, trimmed

2 tablespoons vegetable oil
3 medium baking potatoes,
peeled, cut into quarters
2 large sweet potatoes, peeled,
cut into halves, cut into
quarters
4 small onions, cut into halves

Combine the rosemary, garlic, salt, dry mustard and pepper in a bowl and mix
well. Pat half the seasoning mixture over the surface of the roast. Arrange the
roast fat side up on a rack in a shallow roasting pan. Insert an ovenproof meat
thermometer in the thickest portion of the roast, not resting in the fat.

Add the oil to the remaining seasoning mixture and mix well. Add the
baking potatoes, sweet potatoes and onions to the oil mixture and toss to
coat. Arrange the vegetables around the roast. Do not cover or add water.
Roast at 350 degrees for 1³/4 hours for medium-rare or until the meat
thermometer registers 135 degrees. For medium, roast for 2 hours or until the
meat thermometer registers 150 degrees. Roast the vegetables for 1¹/2 hours
or until tender. Let stand for 15 minutes. Slice as desired. Serve the roast with
the vegetables. You may substitute 2 teaspoons crushed dried rosemary leaves
for the fresh rosemary.

Serves 8

Slow-Cooker Roast

1 (4-pound) beef or pork roast
1 (26-ounce) can cream of
mushroom soup

1 envelope onion soup mix
1/3 cup water

Place the roast in a slow cooker. Pour the soup over the roast. Sprinkle the soup mix around the roast and on top of the roast. Add the water.

Cook, covered, on Low for 8 hours; turn the roast. Cook for 30 to 60 minutes longer or until the desired degree of doneness. Serve the roast and gravy with hot cooked rice.

Serves 8

Baked Brisket

2 large onions, sliced
Paprika to taste
Seasoned salt to taste

1 (4-pound) beef brisket
1 (12-ounce) can Coca-Cola

Arrange the onion slices over the bottom of a roasting pan. Sprinkle with paprika and seasoned salt. Place the brisket fat side up on top of the onions. Sprinkle with paprika and seasoned salt.

Bake, uncovered, at 450 to 500 degrees for 30 minutes. Pour the Coca-Cola over the brisket. Reduce the heat to 325 degrees. Bake, covered with foil, for 2 1/2 to 3 hours or until tender, basting occasionally with the pan juices and adding additional Coca-Cola if desired.

Serves 8

Season the brisket the night before cooking it for the best flavor. Short and fat briskets with a thick covering of fat have the best flavor.

Beef and Beer Buffet

2 pounds round steak
1/4 cup flour
1 envelope onion soup mix
1 (12-ounce) can beer

Cut the round steak into 1-inch cubes and trim any excess fat. Place the steak cubes in a sealable plastic bag. Add the flour and seal tightly. Toss to coat.

Place the steak in a slow cooker. Sprinkle with the soup mix and pour the beer over the top. Cook, covered, on Low for 8 hours. Serve over hot cooked rice or noodles.

Serves 6

Beef is an excellent source of protein and iron, in addition to supplying B-12, zinc, and niacin. It is important to know that iron is the single nutrient most often deficient in the diets of American women. The American Heart Association states that red meat can be eaten three times a week. Today's beef does have a place in a healthy diet!

Chinese Pepper Steak

1 1/2 pounds sirloin steak,
1 inch thick
4 garlic cloves, crushed
1/4 cup vegetable oil
1 teaspoon salt
1 teaspoon ginger
Pepper to taste
1 tablespoon cornstarch
1/4 cup water
3 large green bell peppers,
sliced

2 large onions, thinly sliced
1 (8-ounce) can water
chestnuts, drained, sliced
1/2 cup beef bouillon
1/4 cup soy sauce
4 green onions, cut into
1-inch pieces
1/2 teaspoon sugar

Freeze the steak for 1 hour or longer. Cut into 1/8-inch slices. Sauté the garlic in the oil in a skillet until golden brown. Stir in the salt, ginger and pepper. Add the steak and mix well. Sauté for 2 minutes. Remove the steak to a platter using a slotted spoon, reserving the pan drippings.

Dissolve the cornstarch in the water in a small bowl. Add the bell peppers and sliced onions to the reserved pan drippings. Cook for 3 minutes, stirring frequently. Return the steak to the skillet. Stir in the cornstarch mixture, water chestnuts, bouillon, soy sauce, green onions and sugar. Simmer for 2 minutes or until thickened, stirring constantly. Serve over hot cooked rice.

Serves 6

Shoppers should use labeling information and observation to choose meats. Beef consumers are provided with various kinds of information on the fresh beef sold in grocery stores, including quality grade, species of meat, standardized and common cut names, brand or store label, price per unit, weight, total package costs, refrigeration and cooking suggestions, packaging date, and limited nutrient information.

Rosemary Pepper Beef Steaks

1 tablespoon olive oil
2 teaspoons finely chopped fresh rosemary
2 large garlic cloves, crushed
3/4 teaspoon coarsely ground pepper
4 boneless beef top loin steaks, 1 inch thick

Combine the olive oil, rosemary, garlic and pepper in a bowl and mix well. Press the olive oil mixture evenly into both sides of the steaks. Arrange the steaks in the center of a grill rack over medium-hot coals.

Grill for 15 to 18 minutes for medium-rare to medium or until of the desired degree of doneness, turning occasionally.

Serves 4

Slow-Cooker Beef Stew

2 pounds beef stew meat
1 cup water
3 onions, cut into quarters
3 carrots, coarsely chopped
1 garlic clove, minced
1 rib celery, chopped
3 potatoes, peeled, coarsely chopped
Salt and pepper to taste
1 teaspoon paprika
1 teaspoon Worcestershire sauce
1 bay leaf

Cut the stew meat into 1^1/2-inch cubes. Layer the stew meat, water, onions, carrots, garlic, celery, potatoes, salt, pepper, paprika, Worcestershire sauce and bay leaf in the order listed in a 5-quart slow cooker and stir just enough to mix the seasonings. Cook, covered, on Low for 10 to 12 hours, stirring occasionally. Discard the bay leaf. Spoon into bowls.

Serves 6

Beer Stew

2 pounds beef stew meat
Salt and pepper to taste
1/2 cup flour
1/2 cup vegetable oil
2 (12-ounce) cans beer
4 or 5 red potatoes, peeled,
  cut into chunks
3 or 4 carrots, peeled,
  cut into chunks

1 onion, chopped
2 garlic cloves, chopped
2 tablespoons soy sauce
2 tablespoons Worcestershire
  sauce
2 bay leaves
1 teaspoon thyme

Sprinkle the stew meat with salt and pepper and coat with the flour. Brown the stew meat in the oil in a skillet. Drain on paper towels.

Combine the stew meat, beer, potatoes, carrots, onion, garlic, soy sauce, Worcestershire sauce, bay leaves and thyme in a stockpot and mix well. Simmer for 3 to 4 hours or until of the desired consistency, stirring occasionally. Discard the bay leaves. You may cook in a slow cooker for 8 to 10 hours. Substitute venison for the beef if desired.

Serves 6

Stuffed Bell Peppers

6 large bell peppers
1 tablespoon olive oil
8 ounces lean ground beef
1 cup chopped onion
1 cup chopped fresh tomato
1/2 cup chopped green onions
1/2 cup chopped fresh parsley
2 teaspoons chopped garlic

2 cups cooked rice
Salt to taste
Louisiana hot sauce or
  cayenne pepper to taste
1 (8-ounce) can tomato sauce
1/2 cup dry white wine
1/2 cup water

Slice the tops from the bell peppers. Discard the membranes and seeds. Rinse the bell peppers inside and outside and drain.

Heat the olive oil in a skillet over medium-high heat. Brown the ground beef in the hot olive oil, stirring until crumbly; drain. Stir in the onion, tomato, green onions, parsley and garlic. Cook until the onions are tender, stirring frequently. Stir in the rice, salt and hot sauce. Remove from heat.

Spoon some of the ground beef mixture into each bell pepper. Stand the bell peppers upright in a baking dish. Mix the tomato sauce, wine and water in a bowl and pour around the bell peppers, not on top. Bake at 350 degrees for 1 hour.

Serves 6

Eggplant-Stuffed Bell Peppers

4 large bell peppers
1 medium onion, chopped
1 garlic clove, minced
1 tablespoon chopped celery
2 tablespoons vegetable oil or margarine
2 medium eggplant, peeled, chopped

1 medium tomato, peeled, chopped
1 pound ground beef, browned, drained
1 cup cooked rice
Salt and pepper to taste
1/2 cup bread crumbs

Cut the bell peppers horizontally into halves. Discard the membranes and seeds. Sauté the onion, garlic and celery in the oil in a skillet. Stir in the eggplant and tomato.

Cook over medium heat until the vegetables are tender, stirring frequently. Add the ground beef and rice and mix well. Season with salt and pepper. Spoon some of the ground beef mixture into each bell pepper half. Sprinkle with the bread crumbs. Arrange the bell peppers stuffing side up in a baking dish. Bake at 350 degrees for 30 minutes or until the bell peppers are tender.

Serves 4

Bell peppers are grown extensively throughout southeastern Louisiana, with production concentrated in Tangipahoa and surrounding parishes. The Hammond Research Station conducts field studies to develop the most efficient and effective methods for bell pepper production.

Lasagna

1/2 cup chopped onion
2 tablespoons vegetable oil
1 pound ground beef
3 tablespoons parsley flakes
1 teaspoon garlic powder
1 teaspoon salt
1/4 teaspoon pepper
1 (15-ounce) can tomato
    sauce

1 (10-ounce) can tomato soup
8 ounces lasagna noodles
1 pound mozzarella cheese,
    shredded
1/2 cup grated Parmesan cheese
2 eggs, beaten
2 tablespoons parsley flakes
Salt and pepper to taste
1/4 cup grated Parmesan cheese

Sauté the onion in the oil in a skillet until tender. Add the ground beef and mix well. Cook until the ground beef is brown and crumbly, stirring frequently; drain. Stir in 3 tablespoons parsley flakes, garlic powder, salt and pepper.

Simmer for 1 minute, stirring frequently. Stir in the tomato sauce and soup. Simmer while preparing the remaining ingredients, stirring occasionally. Cook the pasta using package directions; drain.

Combine the mozzarella cheese, 1/2 cup Parmesan cheese, eggs, 2 tablespoons parsley flakes, salt and pepper in a bowl and mix well. Spread a thin layer of the ground beef mixture over the bottom of a 3-quart baking dish. Layer with 1/2 of the pasta, 1/2 of the cheese mixture and 1/2 of the remaining ground beef mixture. Top with the remaining pasta, remaining cheese mixture and remaining ground beef mixture. Sprinkle with 1/4 cup Parmesan cheese. Bake at 350 degrees for 30 minutes.

Serves 8

No-Boil Easy Lasagna

2 pounds lean ground chuck
2 medium onions, chopped
2 green onions, chopped
1 rib celery, chopped
1/2 bell pepper, chopped
2 garlic cloves, chopped
2 (25-ounce) cans any flavor
   spaghetti sauce

1 (15-ounce) can Italian-style
   tomatoes
1/2 cup water
16 ounces lasagna noodles
2 pounds mozzarella cheese,
   shredded
1 (7-ounce) can grated
   Parmesan cheese

Brown the ground chuck in a skillet, stirring until crumbly; drain. Stir in the onions, green onions, celery, bell pepper and garlic. Cook until the vegetables are tender, stirring frequently. Add the spaghetti sauce and undrained tomatoes and mix well.

Simmer for 30 minutes or longer, stirring occasionally. Add the water to the sauce at the end of the simmering process; the sauce will be very thin.

Layer the uncooked noodles, sauce mixture, mozzarella cheese and Parmesan cheese 1/3 at a time in a 9×13-inch baking dish. Bake at 350 degrees for 30 minutes. The noodles will absorb the excess water and soften. Serve with hot crusty garlic bread.

Serves 12

Creole Meatballs

Creole Sauce

2 tablespoons vegetable oil
2 tablespoons flour
1 cup (about) water
1 (15-ounce) can stewed
tomatoes
1 (8-ounce) can tomato
sauce
1/4 cup chopped green
onions

1/4 cup chopped bell pepper
1/4 cup chopped
fresh parsley
2 tablespoons lemon juice
1 teaspoon grated
lemon zest
1/2 teaspoon finely
chopped garlic

Salt and cayenne pepper
to taste
2 tablespoons bacon
drippings
1/4 cup heavy dark
steak sauce
2 tablespoons sugar

Meatballs

11/2 pounds ground round
1/2 cup bread crumbs
1 egg, lightly beaten
1 small onion, chopped
1/4 cup chopped fresh parsley

1 teaspoon salt
1 teaspoon basil
1 teaspoon thyme
Cayenne pepper to taste
2 tablespoons vegetable oil

For the sauce, heat the oil in a cast-iron Dutch oven. Add the flour gradually, stirring constantly. Cook until dark brown, stirring constantly. Add the water, undrained tomatoes and tomato sauce; mash the tomatoes.

Sauté the green onions, bell pepper, parsley, lemon juice, lemon zest, garlic, salt and cayenne pepper in the bacon drippings in a skillet. Add the green onion mixture to the tomato mixture and mix well. Stir in the steak sauce and sugar. Cook for 15 to 20 minutes or until thickened, stirring constantly and adding additional water if needed for the desired consistency.

For the meatballs, combine the ground round, bread crumbs and egg in a bowl and mix well. Add the onion, parsley, salt, basil, thyme and cayenne pepper and mix just until blended. Shape the ground round mixture into 1-inch balls by hand or with a meat baller, rinsing between each meatball. Fry the meatballs in the oil in a skillet until brown on all sides; drain.

Add the meatballs to the sauce and mix gently. Simmer for 30 to 60 minutes, stirring occasionally. Serve over hot cooked spaghetti.

Serves 8

The Kids Will Never Know Poorboys

1 large loaf French bread
1/4 cup mayonnaise
1/2 cup finely chopped
  broccoli florets
1/2 cup finely chopped
  cauliflower florets
8 ounces ground beef
8 ounces breakfast sausage

Salt and pepper to taste
1 cup chopped onion
1/2 cup sliced olives
1/2 cup sliced mushrooms
3/4 cup shredded jalapeño
  cheese
3/4 cup shredded Monterey
  Jack cheese

Split the bread loaf horizontally into halves. Spread the cut sides with the mayonnaise. Layer both halves evenly with the broccoli and cauliflower. Brown the ground beef and sausage in a skillet, stirring until crumbly; drain. Season with salt and pepper.

Spoon the ground beef mixture over the prepared layers. Top with the onion, olives and mushrooms. Sprinkle with the jalapeño cheese and Monterey Jack cheese.

Arrange the bread halves on a baking sheet. Bake at 350 degrees for 25 minutes or until bubbly. Slice as desired.

Serves 6

Po-Boy (poor-boy) is a New Orleans tradition. It is made with a loaf of fresh French bread, sliced into halves lengthwise then filled with meat or seafood, lettuce, tomatoes, onions, and many other ingredients. It is also known as a submarine.

Neapolitan Beef Pie

2 (8-count) cans crescent dinner rolls
1 (8-ounce) can grated Parmesan cheese
1 egg, beaten
1 pound ground beef
1 white onion, chopped
1 (8-ounce) can sliced mushrooms, drained
1 (8-ounce) can tomato sauce
8 ounces Cheddar cheese, shredded

Unroll 1 can of the crescent roll dough and separate into rectangles. Press the rectangles over the bottom of a baking dish, pressing the edges and perforations to seal. Whisk the Parmesan cheese and egg in a bowl. Spread the egg mixture over the prepared layer.

Brown the ground beef with the onion and mushrooms in a skillet, stirring until the ground beef is crumbly; drain. Stir in the tomato sauce. Pour the tomato sauce mixture over the prepared layers. Sprinkle with the Cheddar Cheese.

Unroll the remaining can of crescent roll dough and separate into rectangles. Arrange the rectangles over the top of the prepared layers. Bake at 350 degrees for 1 hour.

Serves 4

South-of-the-Border Bake

1 (8-count) can refrigerated corn bread twists
1 pound lean ground chuck
1 (10-ounce) can cream of chicken soup
4 ounces hot Pepper Jack cheese, shredded
1/4 cup chopped green onions
1 envelope taco seasoning mix
1/2 teaspoon Creole seasoning
1 cup shredded Cheddar cheese
Chopped jalapeño chiles (optional)
Chopped fresh parsley (optional)

Roll the corn twists on a lightly floured surface into a circle large enough to fit a 9-inch pie plate. Fit the pastry into the pie plate, crimping the edges. Prick the bottom and side with a fork. Bake at 350 degrees for 10 minutes.

Brown the ground chuck in a skillet, stirring until crumbly; drain. Stir in the soup, Pepper Jack cheese, green onions, seasoning mix and Creole seasoning. Spoon the ground beef mixture into the prepared pie plate. Sprinkle with the Cheddar cheese. Bake for 20 minutes. Garnish each serving with chopped jalapeño chiles and chopped fresh parsley.

Serves 6

"More" Casserole

1 pound ground beef
1 medium onion, chopped
2 garlic cloves, chopped
1 (6-ounce) can tomato paste
Sugar to taste
1 (17-ounce) can
  cream-style corn

1 tomato paste can water
Salt and pepper to taste
8 ounces medium noodles
1 cup shredded Cheddar
  cheese

Brown the ground beef with the onion and garlic in a skillet, stirring until the ground beef is crumbly; drain. Stir in the tomato paste. Cook until thickened, stirring frequently. Stir in the sugar. Add the corn, water, salt and pepper and mix well.

Cook the noodles using package directions; drain. Layer 1/2 of the ground beef mixture, noodles and remaining ground beef mixture in a baking dish. Sprinkle with the cheese. Bake at 350 degrees for 20 to 25 minutes or until brown and bubbly.

Serves 4

Roast Pork

2 tablespoons butter or margarine
2 cups fresh or frozen orange juice
1 tablespoon grated orange zest
1 tablespoon salt
1 tablespoon black pepper
2 garlic cloves, minced
1/8 teaspoon each cayenne pepper and oregano
1 (6-pound) pork loin roast

Heat the butter in a saucepan. Stir in the orange juice, orange zest, 1 teaspoon of the salt, 1/2 teaspoon of the black pepper, garlic, cayenne pepper and oregano.

Pat the remaining 2 teaspoons salt and remaining 2 1/2 teaspoons black pepper over the surface of the pork. Arrange the pork in a roasting pan. Drizzle with the butter mixture.

Roast at 350 degrees for 3 hours or until brown and cooked through, basting frequently. Remove the pork to a platter and slice as desired. Serve with the pan drippings.

Serves 10

Pork and Cabbage Casserole

1 large head cabbage, sliced
2 to 3 pounds ground pork
1/2 cup (1 stick) butter
2 bunches shallots with tops, chopped
2 onions, chopped
2 green bell peppers, chopped
Chopped garlic to taste
1 or 2 (10-ounce) cans cream of chicken soup
Crushed red pepper to taste
Salt to taste
4 to 5 cups cooked white rice

Steam the cabbage until tender-crisp; drain. Brown the ground pork in the butter in a large skillet, stirring until the ground pork is crumbly; drain. Stir in the shallots, onions, bell peppers and garlic. Cook for 15 minutes, stirring occasionally. Stir in the soup, red pepper and salt.

Cook for 10 minutes, stirring frequently. Add the cabbage and rice to the ground pork mixture and mix well. Spoon the pork mixture into a buttered 9×13-inch baking dish. Bake at 350 degrees for 15 minutes. May be prepared in advance and stored, covered, in the freezer for future use. Bake just before serving.

Serves 12

Savory Ham Pie

3 tablespoons chopped onion
1/4 cup chopped green
bell pepper
1/4 cup vegetable oil
6 tablespoons flour
1 1/3 cups milk

1 (10-ounce) can cream of
chicken or cream of
mushroom soup
1 1/2 cups chopped cooked ham
1 tablespoon lemon juice
1 recipe cheese biscuit dough

Sauté the onion and bell pepper in the oil in a skillet. Stir in the flour. Add the milk and soup and mix well. Cook until thickened, stirring frequently. Stir in the ham and lemon juice. Spoon the ham mixture into a greased baking dish.

Roll the biscuit dough on a lightly floured surface and cut with a round cutter. Arrange the rounds over the top of the prepared layer. Bake at 425 degrees for 15 minutes. Reduce the oven temperature to 400 degrees. Bake for 10 minutes longer. You may bake the biscuits separately and place on the ham pie just before serving. Substitute plain or angel biscuits for the cheese biscuits if desired, sprinkling shredded cheese over the top of the pie just before the end of the baking process.

Serves 6

Chicken Spaghetti

1 (3-pound) chicken
10 ounces spaghetti
1 cup chopped onion
1 cup chopped bell pepper
1 cup chopped celery

½ cup (1 stick) margarine
1 (10-ounce) can cream of
    mushroom soup
1 (10-ounce) can tomato soup
1 tablespoon chili powder

Combine the chicken with enough water to cover in a stockpot. Bring to a boil. Boil until the chicken is tender. Drain, reserving 2 cups of the broth. Chop the chicken into bite-size pieces, discarding the skin and bones. Cook the spaghetti using package directions; drain.

Sauté the onion, bell pepper and celery in the margarine in a skillet until tender. Stir in the reserved broth, chicken, soups and chili powder. Simmer for 30 minutes, stirring occasionally. Add the pasta and mix well.

Spoon the chicken mixture into a 9×13-inch baking dish. Bake at 350 degrees for 25 minutes. You may sprinkle with shredded cheese if desired.

Serves 8

Through a new youth initiative called Character Counts, Louisiana 4-H is taking the six pillars of character—trustworthiness, respect, responsibility, fairness, caring, and citizenship—to schools and other organizations. School administrators and teachers are finding better behaved students and fewer discipline problems.

Chicken and Corn Bread Dressing

1 (3½-pound) chicken, cut up
1 large onion, finely chopped
5 green onions, finely chopped
2 ribs celery, finely chopped
2 tablespoons margarine
1 recipe corn bread, baked, cooled

2 or 3 slices dry white bread
1 cup cold cooked rice
3 eggs, lightly beaten
Sage to taste
Poultry seasoning to taste
Salt and pepper to taste

Combine the chicken with enough water to cover in a stockpot. Bring to a boil. Boil until the chicken is tender. Drain, reserving the broth.

Sauté the onion, green onions and celery in the margarine in a skillet until tender but not brown. Crumble the corn bread and white bread into a bowl and mix well. Stir in the rice. Add the sautéed onion mixture, eggs, sage, poultry seasoning, salt and pepper and mix well. Add enough of the reserved broth to make a soupy mixture and mix well.

Spoon the corn bread mixture into a 9×13-inch baking pan. Arrange the chicken over the prepared layer. Bake at 350 degrees for 45 minutes.

Serves 8

Old-Fashioned Chicken and Dumplings

1 (3-pound) chicken
Salt and pepper to taste
3 cups flour
1 teaspoon salt
1/2 teaspoon baking powder
2 tablespoons vegetable oil

1 (10-ounce) can cream of
  chicken soup
Poultry seasoning to taste
1/4 cup (1/2 stick) margarine, or
  to taste

Combine the chicken and salt and pepper to taste with enough water to cover in a stockpot. Bring to a boil. Boil until the chicken is tender. Drain, reserving the broth. Cut the chicken into bite-size pieces, discarding the skin and bones.

Sift the flour, 1 teaspoon salt and baking powder into a bowl and mix well. Pour the oil into a measuring cup. Add enough water to the measuring cup to measure 1 cup. Add the water mixture to the flour mixture and mix until of the consistency of a biscuit dough. Roll the dough very thin on a heavily floured surface. Cut the dough into 3-inch squares.

Bring the reserved broth to a rolling boil. Stir in the soup and poultry seasoning. Drop the dough squares into the boiling broth 1 at a time. Reduce the heat to low.

Simmer, covered, for 20 minutes; do not remove the cover during the cooking process. Stir in the chicken and margarine.

Serves 6

Easy Chicken and Dumplings

1 (3-pound) chicken
1 (14-ounce) can chicken broth
1 (10-count) package flour tortillas
Salt and pepper to taste

Combine the chicken with enough water to cover in a stockpot. Bring to a boil. Boil until the chicken is tender. Remove the chicken to a platter, reserving the broth. Chill the broth in the refrigerator and skim the fat. Chop the chicken, discarding the skin and bones.

Bring the reserved broth and canned broth to a boil in the stockpot. Cut the tortillas into strips. Add to the boiling broth a few at a time. Cook for 10 minutes. Reduce the heat. Stir in the chicken. Cook just until heated through. Season with salt and pepper.

Serves 8

Chicken and Gravy

2 tablespoons flour
1/2 teaspoon each garlic powder and salt
1/4 teaspoon pepper
1/2 teaspoon Cajun seasoned salt
1 (3- to 4-pound) chicken, cut up, trimmed
2 tablespoons vegetable oil
1 onion, chopped
1 1/2 cups water
1/8 teaspoon Worcestershire sauce
8 ounces fresh mushrooms, sliced

Mix the flour, garlic powder, salt, pepper and Cajun seasoned salt in a sealable plastic bag. Add the chicken, seal tightly and shake until the chicken is coated.

Heat the oil in a Dutch oven. Brown the chicken on all sides in the hot oil. Stir in the onion. Add the water, Worcestershire sauce and mushrooms and mix well. Cook, covered, over low heat for 1 1/2 hours, stirring occasionally. Serve with mashed potatoes.

Serves 6

Poultry is the largest animal commodity in the state of Louisiana, surpassing the income generated by beef, dairy, and pork combined. In addition, the poultry industry is growing at about five percent per year, both in Louisiana and in the United States. The success of the poultry industry has resulted from the development of more efficient production systems and increased consumer demand for an economical and nutritious food product.

Chicken Pie

1 (2½- to 3-pound) chicken
2 cups flour
1 teaspoon salt
⅔ cup shortening
¼ cup (½ stick) butter
1 to 1½ cups milk
1 (10-ounce) can cream of chicken soup
Salt and pepper to taste

Combine the chicken with enough water to cover in a stockpot. Bring to a boil. Boil until tender. Drain, reserving the broth. Chop the chicken into bite-size pieces, discarding the skin and bones.

Combine the flour and salt in a bowl and mix well. Cut in the shortening until crumbly. Add just enough ice water to form an easily handled dough and mix well. Roll half the dough ¼ inch thick on a lightly floured surface. Cut into 1×1½-inch strips.

Bring the reserved broth to a boil. Stir in the chicken, butter, milk, soup, salt and pepper. Add the dough strips 1 at a time. Cook for 6 minutes or until the dumplings are tender, stirring occasionally. Pour the dumpling mixture into a 2½- to 3-quart baking dish.

Roll the remaining dough on a lightly floured surface to fit the top of the baking dish. Fit the dough over the top of the dumpling mixture. Crimp the edge and make slits in the top. Bake at 400 degrees for 30 minutes or until golden brown.

Serves 6

Oven-Barbecued Chicken

1 cup ketchup
½ cup water
¼ cup wine vinegar
¼ cup chopped onion
¼ cup chopped green bell pepper
1½ tablespoons Worcestershire sauce
1 tablespoon prepared mustard
1 tablespoon brown sugar
4 chicken breasts, or 1 small chicken, cut up
1 tablespoon garlic salt, or to taste
¼ teaspoon pepper

Combine the ketchup, water, wine vinegar, onion, bell pepper, Worcestershire sauce, prepared mustard and brown sugar in a saucepan and mix well. Bring to a boil; reduce the heat.

Simmer over low heat for 5 minutes, stirring occasionally. Remove from heat. Cover to keep warm.

Sprinkle the chicken with the garlic salt and pepper. Arrange the chicken in a single layer in a baking dish. Pour the barbecue sauce over the chicken. Bake, covered, at 325 degrees for 1 hour or until cooked through.

Serves 4

New Orleans Chicken

1/2 cup chopped onion
1/2 cup chopped celery
1/2 cup (1 stick) margarine
1 (2 1/2- to 3-pound) chicken, cut up
1 cup water
2 tablespoons chopped fresh parsley
2 small bay leaves
2 garlic cloves, minced
1/2 teaspoon thyme

Sauté the onion and celery in the margarine in a heavy skillet for 3 minutes or until the onion is golden brown. Add the chicken. Cook until brown on both sides, turning once. Stir in the water, parsley, bay leaves, garlic and thyme.

Bring to a boil over high heat; reduce the heat. Simmer for 1 hour or until the chicken is cooked through, stirring occasionally. Discard the bay leaves. Serve the pan drippings with the chicken.

Serves 4 to 6

Chicken and Crawfish

4 boneless skinless chicken breasts
1/2 cup flour
Salt and pepper to taste
1/2 cup (1 stick) butter or margarine
2 cups chopped onions
1 cup chopped celery
1/2 cup chopped bell pepper
6 garlic cloves, minced
1 (15-ounce) can diced tomatoes
1 (14-ounce) can chicken broth
1 (8-ounce) can tomato sauce
1 to 2 pounds peeled crawfish tails

Cut the chicken into bite-size pieces. Mix the flour, salt and pepper in a shallow dish. Coat the chicken on both sides with the flour mixture. Cook the chicken in the butter in a large skillet for 5 minutes, turning frequently. Remove the chicken to a platter with a slotted spoon, reserving the pan drippings.

Add the onions, celery, bell pepper and garlic to the reserved pan drippings. Sauté until the vegetables are tender. Stir in the chicken, undrained tomatoes, broth and tomato sauce. Simmer for 15 minutes, stirring occasionally. Add the crawfish and mix well. Simmer for 30 minutes longer, stirring occasionally. Serve over hot cooked rice.

Serves 8

Easy Chicken Quesadillas

4 boneless skinless chicken breasts
1 medium onion, chopped
1 tablespoon margarine or vegetable oil
Chopped green onions to taste
1 (10-ounce) can cream of mushroom soup
Salt, pepper and favorite seasonings to taste
8 (10-inch) flour tortillas
1 cup shredded Monterey Jack cheese
Sour cream (optional)
Salsa (optional)

Cut the chicken into thin strips. Sauté the onion in the margarine in a nonstick skillet until tender. Add the chicken and green onions. Sauté for 10 minutes. Stir in the soup, salt, pepper and favorite seasonings. Cook until heated through, stirring frequently.

Spray a skillet with nonstick cooking spray. Place 1 tortilla in the prepared skillet over medium heat. Spread the tortilla with some of the chicken mixture and sprinkle with 2 tablespoons of the cheese. Top with another tortilla.

Cook for 1 minute on each side or until the cheese melts. Remove the quesadilla to a platter. Repeat the process with the remaining tortillas, chicken mixture and cheese. Cut each quesadilla into wedges and serve hot with sour cream and salsa. You may substitute 1 pound crawfish tails for the chicken.

Serves 4 to 8

Baked Chicken Parmesan

2 cups fine bread crumbs
1/2 cup grated Parmesan cheese
1/3 cup chopped fresh parsley
1 (3-ounce) can French-fried onions, crushed
2 garlic cloves, crushed
1/4 cup (1/2 stick) margarine
1 teaspoon Worcestershire sauce
1 teaspoon dry mustard
8 boneless skinless chicken breasts

Line a 9×13-inch baking dish with foil. Combine the bread crumbs, cheese, parsley and onions in a bowl and mix well. Sauté the garlic in the margarine in a saucepan over medium-high heat for 1 minute. Remove from heat. Stir in the Worcestershire sauce and dry mustard.

Dip the chicken in the margarine mixture and coat with the crumb mixture. Arrange the chicken in a single layer in the prepared baking dish. Drizzle with the remaining butter mixture. Bake at 350 degrees for 50 to 60 minutes or until the chicken is cooked through.

Serves 8

Italian Chicken

3 ounces cream cheese,
softened
1 ounce thinly sliced
prosciutto, chopped
1 teaspoon lemon juice
6 medium chicken breasts
12 to 18 large fresh basil leaves

1/2 teaspoon salt
1/2 teaspoon dried basil
1 teaspoon lemon juice
1/2 teaspoon flour
1/2 teaspoon chicken bouillon
granules

Combine the cream cheese, prosciutto and 1 teaspoon lemon juice in a bowl and mix well. Loosen the skin carefully from each chicken breast to form a pocket. Place 2 or 3 basil leaves and 1/6 of the cream cheese mixture in each pocket.

Arrange the chicken skin side up in a roasting pan. Sprinkle with the salt and 1/2 teaspoon dried basil. Bake at 400 degrees for 35 to 45 minutes or until the juices run clear when the chicken is pierced with a sharp knife, basting occasionally with the pan drippings. Remove the chicken to a platter, reserving the pan drippings. Cover the chicken to keep warm.

Pour the reserved pan drippings into a measuring cup. Let stand until cool. Skim to remove the fat. Add enough water to the reserved pan drippings to measure 3/4 cup. Stir in 1 teaspoon lemon juice. Add the flour and bouillon granules and mix well. Pour into the roasting pan and stir to loosen any browned bits.

Cook over medium heat until thickened, stirring constantly. Drizzle the sauce over the chicken. Garnish with additional fresh basil leaves. Serve immediately.

Serves 6

Stuffed Italian Chicken Rolls

6 large boneless skinless
chicken breasts
1/4 teaspoon salt
1/4 teaspoon pepper
3 ounces reduced-fat cream
cheese, softened

1/4 cup Italian salad dressing
1/2 cup finely chopped red
bell pepper
2 tablespoons parsley flakes
3/4 cup cornflake crumbs
1/2 teaspoon paprika

Pound the chicken 1/4 inch thick between sheets of plastic wrap using a meat mallet or rolling pin. Sprinkle the chicken with salt and pepper.

Combine the cream cheese, salad dressing, bell pepper and parsley flakes in a bowl and mix well. Spread 2 tablespoons of the cream cheese mixture over each chicken fillet. Roll to enclose the filling and secure with wooden picks.

Combine the cornflake crumbs and paprika in a bowl and mix well. Coat the chicken rolls with the crumb mixture. Arrange the rolls in a 7×11-inch baking dish sprayed with nonstick cooking spray. Chill, covered, for 8 hours.

Let the chicken rolls stand at room temperature for 30 minutes. Bake at 350 degrees for 35 minutes or until cooked through. Let stand for 10 minutes. Discard the wooden picks and cut into 1-inch slices with an electric knife.

Serves 6

Chicken and Angel Hair Pasta

2 boneless skinless chicken breasts	1 (10-ounce) package frozen broccoli florets, thawed
Salt and pepper to taste	2 garlic cloves, minced
2 tablespoons olive oil	2/3 cup chicken broth
12 ounces angel hair pasta	1/4 cup grated Parmesan cheese
1 carrot, sliced diagonally	1 teaspoon basil

Cut the chicken into bite-size pieces. Sprinkle the chicken with salt and pepper. Heat 1 tablespoon of the olive oil in a skillet over medium heat. Add the chicken. Cook for 5 minutes or until the chicken is cooked through, turning several times. Remove the chicken to a platter with a slotted spoon, reserving the pan drippings.

Heat the reserved pan drippings with the remaining 1 tablespoon olive oil. Stir in the carrot. Cook for 4 minutes, stirring constantly. Stir in the broccoli and garlic. Cook for 2 minutes, stirring constantly. Cook the pasta in a large saucepan using package directions; drain.

Add the chicken, broth, cheese and basil to the carrot mixture. Simmer for 4 minutes, stirring occasionally. Spoon the chicken mixture over the pasta on a serving platter.

Serves 6

The popularity of pasta is at an all-time high, primarily because people have recognized it as a versatile, low-fat source of complex carbohydrates. Pasta provides valuable amounts of carbohydrates, protein, B vitamins, and iron. Egg-free pasta contains no cholesterol and pasta with eggs is low in cholesterol.

Quick Chicken Jambalaya

6 boneless skinless chicken thighs
2 sausage links, sliced
1 cup (2 sticks) butter or margarine
1 onion, chopped
1 bell pepper, chopped
1/2 cup chopped green onions
2 cups cooked rice
Cajun seasoned salt to taste

Cut the chicken into bite-size pieces. Brown the chicken and sausage in the butter in a cast-iron skillet, stirring frequently. Remove the chicken mixture to a bowl with a slotted spoon, reserving the pan drippings.

Sauté the onion, bell pepper and green onions in the reserved pan drippings until tender. Add the chicken mixture to the vegetable mixture and mix well. Stir in the rice and Cajun seasoned salt. Cook just until heated through, stirring frequently.

Serves 6

Chicken Corn Bread Bake

4 cups crumbled corn bread
1/4 cup chopped green bell pepper
1/4 cup chopped celery
1/4 cup chopped green onions
1 1/2 cups coarsely chopped cooked chicken breasts (about 3 large chicken breasts)
1 (10-ounce) can cream of chicken soup
1 1/2 cups chicken broth
Salt and pepper to taste

Grease or spray a 9×13-inch baking dish with nonstick cooking spray. Combine the corn bread, bell pepper, celery and green onions in a bowl and mix well. Spoon half the corn bread mixture into the prepared baking dish. Top with the chicken.

Whisk the soup, broth, salt and pepper in a bowl. Pour over the prepared layers; the mixture should have a soupy consistency. Additional broth may be added at this time if not of the desired consistency. Sprinkle with the remaining corn bread mixture and press lightly. Let stand for 20 minutes. Bake at 350 degrees for 45 minutes or until brown and bubbly.

Serves 8

Chicken Enchiladas

1 (16-ounce) can tomatoes
1 (4-ounce) can green chiles
$1/2$ teaspoon coriander seeds
$1/2$ teaspoon salt
1 cup sour cream
2 cups finely chopped cooked chicken
8 ounces cream cheese, softened
$1/4$ cup finely chopped onion
$3/4$ teaspoon salt
2 tablespoons vegetable oil
12 (6-inch) corn tortillas
1 cup shredded Monterey Jack cheese

Combine the undrained tomatoes, undrained green chiles, coriander seeds and $1/2$ teaspoon salt in a blender. Process until smooth. Add the sour cream. Process until blended. Combine the chicken, cream cheese, onion and $3/4$ teaspoon salt in a bowl and mix well.

Heat the oil in a skillet. Dip each tortilla in the hot oil for 10 seconds. Drain on paper towels. Spread some of the chicken mixture in the center of each tortilla. Roll to enclose the filling.

Arrange the tortillas seam side down in a 9×13-inch baking dish sprayed with nonstick cooking spray. Pour the tomato mixture over the top. Bake, covered with foil, at 350 degrees for 30 minutes. Remove the foil and sprinkle with the Monterey Jack cheese. Bake until the cheese melts.

Makes 12 enchiladas

Barbecue Sauce

2 cups vinegar
$1/2$ cup (1 stick) margarine
2 tablespoons hot sauce
2 tablespoons Worcestershire sauce
1 tablespoon paprika
1 tablespoon pepper
1 tablespoon liquid smoke
1 tablespoon salt

Combine the vinegar, margarine, hot sauce, Worcestershire sauce, paprika, pepper, liquid smoke and salt in a saucepan. Bring to a boil; reduce the heat.

Simmer for 20 minutes, stirring occasionally. Use as a basting sauce for chicken.

Makes 11 ($1/4$-cup) servings

Fried Turkey

1 turkey (any size)
1 cup (2 sticks) butter or margarine
1/4 cup Worcestershire sauce
2 tablespoons soy sauce
1/8 teaspoon Tabasco sauce
1/2 bottle garlic juice
Tabasco sauce to taste
Peanut oil for deep-frying

Thaw the turkey using package directions 24 hours prior to cooking. Combine the butter, Worcestershire sauce, soy sauce, 1/8 teaspoon Tabasco sauce and garlic juice in a microwave-safe dish. Microwave until heated through; stir.

Inject the marinade into the meaty portions of the turkey. Rub the outer surface of the turkey generously with Tabasco sauce. Chill, covered with plastic wrap, for 24 hours.

Add enough peanut oil to a turkey fryer to totally submerge the turkey. Heat the peanut oil to 325 degrees. Add the turkey. Fry for 1 1/2 minutes per pound, maintaining the temperature of the peanut oil at 325 degrees. Remove the turkey to a platter. Let stand for 30 minutes before serving.

Variable servings

Pinto Beans with Smoked Turkey

1 pound dried pinto beans
6 cups water
1 (2-pound) smoked turkey leg
1 tablespoon lite salt
1/2 teaspoon pepper
1 envelope saccharin-based artificial sweetener

Sort and rinse the beans. Soak using package directions. Combine the water and turkey in a stockpot. Bring to a boil; reduce the heat. Simmer, covered, for 1 1/2 hours or until the turkey is tender. Add the drained beans, lite salt, pepper and artificial sweetener.

Cook, covered, for 1 1/2 hours or until the beans are tender, stirring occasionally. Remove the turkey to a platter. Chop the turkey, discarding the skin and bones. Return the turkey to the stockpot and mix well. Cook just until heated through. You may substitute any dried beans for the pinto beans.

Serves 8

Stuffed Cornish Game Hen

Potato Stuffing

1 large Irish potato, peeled	2 tablespoons vegetable oil	2 tablespoons chopped
Garlic salt to taste	1/4 onion, chopped	bell pepper
Red pepper to taste		

Cornish Game Hen

1 (1 1/2-pound) Cornish game hen	2 tablespoons chopped bell pepper
Garlic salt to taste	1 small can mushroom sauce
Red pepper to taste	2 tablespoons wine (optional)
2 tablespoons vegetable oil	1/4 (4-ounce) can mushroom stems and
1/4 onion, chopped	pieces, drained

For the stuffing, chop the potato. Season with garlic salt and red pepper. Heat the oil in a skillet; the oil should cover the bottom of the skillet. Add the potato. Cook until tender, stirring frequently. Stir in the onion and bell pepper. Cook until the potatoes are brown, stirring frequently. Let stand until cool.

For the game hen, stuff the game hen with the potato stuffing and truss. Sprinkle the outside surface with garlic salt and red pepper. Heat the oil in a skillet. Brown the game hen in the hot oil. Remove the game hen to a platter, reserving the pan drippings.

Sauté the onion and bell pepper in the reserved pan drippings until the onion is brown. Stir in the mushroom sauce and wine. Cook for 1 to 1 1/2 hours, stirring occasionally and adding water as needed for the desired consistency. Return the game hen to the skillet. Stir in the mushrooms.

Cook until the game hen is cooked through, turning occasionally. Serve the game hen and gravy with hot cooked rice or mashed potatoes.

Serves 1

Cornish game hens are the smallest members of the chicken family.
They are suitable for roasting, especially with a rice stuffing. The average weight
is 1 1/2 pounds or less. Try serving an individual bird to each of your
guests for an elegant holiday meal.

Venison Stew

2 pounds lean venison
Salt and black pepper to taste
Red pepper to taste
1 cup flour
1/4 cup vegetable oil
1 large onion, chopped
3 ribs celery, chopped
3 garlic cloves, chopped
1 cup water

Cut the venison into 4-inch cubes. Sprinkle with salt, black pepper and red pepper. Coat the venison with the flour.

Heat the oil in a Dutch oven over high heat. Brown the venison on all sides in the hot oil. Remove the venison to a bowl using a slotted spoon, reserving the pan drippings. Sauté the onion, celery and garlic in the reserved pan drippings until light brown and tender.

Return the venison to the Dutch oven and mix well. Stir in the water. Cook over medium heat for 1 hour, stirring occasionally. Serve over hot cooked rice.

Serves 6

Venison and Squash

1 pound venison, pork or turkey sausage
1 garlic clove, chopped
4 cups sliced yellow squash
1/2 cup bread crumbs
1/2 cup grated Parmesan cheese
1/2 cup milk
1 teaspoon chopped fresh parsley
1/2 teaspoon oregano
1/2 teaspoon salt
2 eggs, beaten

Brown the sausage with the garlic in a skillet, stirring until the sausage is crumbly; drain. Combine the squash with enough water to cover in a saucepan. Bring to a boil; reduce the heat. Cook until tender; drain.

Combine the sausage, squash, bread crumbs, cheese, milk, parsley, oregano and salt in a bowl and mix well. Fold in the eggs. Spoon the sausage mixture into a 7×11-inch baking dish sprayed with nonstick cooking spray. Bake at 325 degrees for 20 to 25 minutes.

Serves 6

Venison Pot Roast

1 (3- to 4-pound)
venison roast
1/4 cup vegetable oil
Salt and pepper to taste
Garlic cloves to taste

4 cups lemon-lime soda
6 tablespoons ketchup
1/4 cup packed brown sugar
2 tablespoons vinegar

Sear the roast on both sides in the oil in a Dutch oven. Sprinkle with salt and pepper. Make the desired amount of vertical slits in the roast and insert garlic cloves in the slits. Pour the soda over the roast.

Cook, covered, until almost of the desired degree of doneness. Combine the ketchup, brown sugar and vinegar in a bowl and mix well. Pour the ketchup mixture over the roast. Cook, covered, for 30 minutes longer. You may substitute a beef roast for the venison.

Serves 8

Serving Louisiana
Seafood

Catfish Co-Op

12 ounces pasta
1 (16-ounce) package mixed
vegetables (including
broccoli, cauliflower and
carrots), cooked, drained
1 (16-ounce) package mixed
vegetables (including peas,
green beans and carrots),
cooked, drained

1 (10-ounce) can reduced-fat
cream of mushroom soup
1 cup nonfat mayonnaise
2 teaspoons lite seasoned salt
1 teaspoon sugar
1/2 teaspoon thyme
3 pounds catfish fillets,
steamed, cut into bite-size
pieces

Cook the pasta using package directions; drain. Combine the pasta, mixed vegetables, soup, mayonnaise, lite seasoned salt, sugar and thyme in a bowl and mix well. Add the catfish and mix gently.

Spoon the catfish mixture into a microwave-safe dish. Microwave until heated through or bake at 350 degrees for 20 minutes. You may substitute nonfat plain yogurt for the nonfat mayonnaise, 12 ounces pasta sauce for the soup and vegetables of your choice for the mixed vegetables. Prepare in advance and freeze, covered, for future use. Bake just before serving.

Serves 8

Good nutritional advice these days encourages consumers to eat more fish. Fish is an excellent source of high-quality protein, vitamins, and minerals, yet low in fat and calories. When buying fish whole, watch for bright, clear eyes, reddish-pink gills free from odor, and bright-colored scales adhering tightly to skin. The flesh should be firm and elastic.

Catfish à la Creole

2 cups chopped green onions
1 cup chopped fresh parsley
1/2 cup chopped onion
1/2 cup chopped bell pepper
2 tablespoons olive oil
1/2 cup dry white wine
1 tablespoon chopped garlic
3 cups chopped fresh or
   canned tomatoes

1 cup vegetable juice cocktail
   or tomato juice
1 tablespoon Worcestershire
   sauce
Salt to taste
Cayenne pepper or Louisiana
   hot sauce to taste
2 pounds fresh catfish fillets,
   cut into 1-inch pieces

Sauté the green onions, parsley, onion and bell pepper in the olive oil in a skillet over medium-high heat until the green onions and onion are tender. Stir in the wine and garlic. Cook for 10 minutes, stirring frequently. Add the tomatoes, vegetable juice cocktail, Worcestershire sauce, salt and cayenne pepper and mix well. Reduce the heat to medium.

Cook, covered, for 20 minutes longer or until the tomatoes fall apart, stirring occasionally. Add the fish and mix gently. Reduce the heat to low. Simmer for 20 to 30 minutes or until the fish is cooked through, stirring occasionally. Serve over hot cooked rice.

Serves 6

Louisiana is the fourth leading state in the production of catfish. Researchers at the LSU AgCenter's Aquaculture Research Station conduct nutrition research on the development and improvement of feeds for catfish, bass, and tilapia, and improved pond environment. A good pond environment is essential for fast-growing healthy fish. Improved diets offer both economic and environmental benefits to catfish producers.

Crunchy Catfish with Lemon Parsley Sauce

Catfish

2 eggs, beaten
2 tablespoons water
1/2 cup butter cracker crumbs
1/4 cup freshly grated
Parmesan cheese

1 tablespoon Greek seasoning
4 (8-ounce) catfish fillets
2 tablespoons unsalted butter,
melted

Lemon Parsley Sauce

1/2 cup (1 stick) unsalted butter
3 green onions, chopped
2 tablespoons fresh
lemon juice
1 tablespoon parsley flakes

1/2 teaspoon salt
1/2 teaspoon pepper
Worcestershire sauce to taste
Tabasco sauce to taste

For the catfish, whisk the eggs and water in a bowl until blended. Mix the cracker crumbs, cheese and Greek seasoning in a shallow dish. Dip the fillets in the egg mixture and coat with the crumb mixture.

Arrange the fillets in a single layer in a greased baking dish. Drizzle with the butter. Bake at 325 degrees for 30 minutes.

For the sauce, melt the butter in a saucepan. Stir in the green onions, lemon juice, parsley flakes, salt, pepper, Worcestershire sauce and Tabasco sauce. Simmer while the fillets are baking, stirring occasionally. Drizzle over the fillets on a serving platter.

Serves 4

Oven-Fried Catfish

1 pound catfish fillets or steaks
Lite salt and pepper to taste

3/4 cup cornmeal

Cut the catfish into strips or small pieces. Sprinkle with lite salt and pepper. Coat the catfish with the cornmeal and spray all sides with nonstick cooking spray. Arrange the strips 1/8 to 1/2 inch apart in a shallow baking dish.

Bake at 400 to 500 degrees for 20 minutes or until the catfish flakes easily. You may substitute prepared fish meal, branflake crumbs or cornflake crumbs for the cornmeal.

Serves 2

Gaspergou with Clear Sauce

5 pounds cleaned gaspergou
2 large onions, sliced, separated into rings
1 small green or red bell pepper, chopped

1 or 2 garlic cloves, minced
1/2 cup finely chopped green onions
1/2 cup finely chopped parsley
Salt and pepper to taste

Cut the fish across the bone into 2-inch-thick steaks. Spray a 6-quart Dutch oven with nonstick cooking spray. Arrange the fish in a single layer in the prepared Dutch oven. Layer the onions, bell pepper, garlic, green onions, parsley, salt and pepper over the fish. Do not add liquid.

Cook, covered, over medium-low heat for 45 minutes, shaking the Dutch oven every 5 minutes. Do not remove cover during the cooking process. Serve over hot cooked rice.

Serves 6

Salmon Balls

1 (14-ounce) can salmon, tuna or other fish
1 egg, lightly beaten
1/2 cup flour
1/2 medium onion, chopped
1 teaspoon (heaping) baking powder
Peanut oil for deep-frying

Discard the skin and bones from the salmon. Drain the salmon, reserving 1/4 cup of the liquid. Combine the salmon and egg in a bowl and mix well with a fork. Add the flour and onion and stir until mixed. Combine the reserved liquid with the baking powder in a bowl and mix with a fork until the mixture foams. Stir into the salmon mixture.

Heat the oil in a deep skillet. Drop the salmon mixture by teaspoonfuls or tablespoonfuls into the hot oil. Fry until golden brown. Drain on paper towels. Serve hot.

Serves 5

Baked Salmon Croquettes

1 (15-ounce) can pink or red salmon
3/4 cup milk
1/4 cup (1/2 stick) butter or margarine
2 tablespoons finely chopped onion
1/3 cup flour
1/2 teaspoon salt
1/4 teaspoon pepper
1 tablespoon lemon juice
1 cup cornflake crumbs

Drain the salmon, reserving the liquid. Add enough of the milk to the reserved liquid to measure 1 cup. Heat the butter in a heavy saucepan over low heat. Stir in the onion. Cook until tender, stirring frequently. Add the flour gradually, stirring constantly. Cook until bubbly, stirring constantly. Stir in the milk mixture gradually. Cook over medium heat until thickened and bubbly, stirring constantly. Mix in the salt and pepper.

Discard the skin and bones from the salmon and place the salmon in a bowl. Flake with a fork. Stir in the lemon juice, 1/2 cup of the cornflake crumbs and the white sauce. Chill the salmon mixture in the refrigerator.

Shape the salmon mixture into 8 croquettes. Roll in the remaining 1/2 cup cornflake crumbs. Arrange the croquettes on a lightly greased baking sheet. Bake at 400 degrees for 30 minutes.

Makes 8 croquettes

Parmesan Flounder Fillets

2 pounds flounder or sole
fillets, 1/2 inch thick
1 cup mayonnaise
1/4 cup snipped fresh chives
1 egg white, stiffly beaten

2 to 3 tablespoons freshly
grated Parmesan cheese
1 tablespoon minced fresh
parsley
Lemon slices (optional)

Rinse the fillets and pat dry. Arrange in a single layer in a baking dish. Mix the mayonnaise and chives in a bowl. Fold 1/2 to 3/4 of the egg white into the mayonnaise mixture. Do not add too much of the egg white or the mixture will be too thin.

Spread the mayonnaise mixture evenly over the fillets. Sprinkle with the cheese and parsley. Bake at 425 degrees until puffed. Broil for 2 minutes or until golden brown. Garnish each serving with lemon slices. You may substitute any firm white fish for the flounder or sole.

Serves 6

Rinse fish fillets in cold water for several seconds prior to seasoning or dusting with flour. This removes surface bacteria before the fish is cooked.

Deviled Crab

2 tablespoons plus 1 teaspoon Chef Paul Prudhomme's® Seafood Magic®	1 1/2 cups chopped onions	4 egg yolks
1 teaspoon dry mustard	1 cup chopped celery	1 pound lump crab meat, shells and cartilage removed
1/4 teaspoon allspice	2 3/4 cups seafood stock	
1/4 cup (1/2 stick) unsalted butter	1 tablespoon flour	6 tablespoons bread crumbs
	2 tablespoons unsalted butter	3 tablespoons unsalted butter
	1 cup chopped green onions	

Mix the first 3 ingredients in a small bowl. Heat 1/4 cup butter in a skillet over high heat until the butter sizzles. Add the onions and celery. Cook for 4 to 5 minutes or until the vegetables just begin to brown, stirring once. Stir in the seasoning mix. Cook for 3 to 4 minutes or until the mixture sticks to the bottom of the skillet. Add 1/2 cup of the stock. Scrape the bottom of the skillet to dislodge any browned bits. Cook for 2 minutes. Stir in the flour. Cook for 2 minutes or until the mixture sticks to the bottom of the skillet, scraping the bottom whenever a crust forms. Add 1 cup of the stock and stir to dislodge the brown crust. Bring to a simmer.

Simmer for 3 to 5 minutes or until thickened, stirring frequently. Stir in the remaining 1 1/4 cups stock. Cook for 5 minutes, stirring frequently. Add 2 tablespoons butter and the green onions and mix well. Cook for 3 minutes, stirring frequently. Remove from heat. Whisk the egg yolks lightly in a bowl. Whisk 3 or 4 spoonfuls of the hot mixture into the egg yolks. Add the egg yolk mixture to the skillet, whisking constantly. Stir in the crab meat. Spoon about 2/3 cup of the crab meat mixture into each of 6 scallop shells. Top each with 1 tablespoon of the bread crumbs and dot each with 1/2 tablespoon of the butter. Arrange the shells on a baking sheet. Bake at 400 degrees for 20 minutes or until golden brown.

Serves 6

Deviled dishes, mentioned in print as early as 1786, became popular in the nineteenth century. Although the word 'deviled' originally indicated a dish that was fiery because it was prepared with hot seasonings, today deviled foods aren't necessarily hot—merely spicy. Deviled crabs are made almost everywhere in the South, but are also enjoyed from the Gulf Coast to the Chesapeake Bay.
—*Chef Paul Prudhomme*

Stuffed Crabs

1 large onion, finely chopped
1/4 cup finely chopped
bell pepper
2 ribs celery, finely chopped
1/4 cup (1/2 stick) margarine
2 cups crab meat
2 cups cooked rice

1 cup seasoned bread crumbs
1/2 cup mayonnaise
1 teaspoon lemon juice
1 egg, lightly beaten
Salt and pepper to taste
1/4 cup bread crumbs

Sauté the onion, bell pepper and celery in the margarine in a heavy skillet until tender. Stir in the crab meat. Cook over low heat for 3 minutes, stirring occasionally. Remove from heat.

Add the rice to the crab mixture and mix well. Stir in 1 cup bread crumbs, mayonnaise, lemon juice and egg. Season with salt and pepper. Spoon the crab meat mixture into 24 crab shells or ramekins and arrange on a baking sheet. Bake at 350 degrees for 20 minutes. You may bake in a 9×13-inch baking dish.

Serves 12

Boiled Crabs

6 quarts water	2 lemons, cut into quarters
1/3 cup salt	1/4 cup vinegar
2 (3-ounce) bags shrimp and crab boil	Cayenne pepper to taste
	1 head garlic
1 onion, cut into halves	2 dozen live blue crabs, rinsed

Bring the water to a boil in a stockpot. Add the salt, shrimp and crab boil, onion, lemons, vinegar, cayenne pepper and garlic. Boil for 5 minutes. Add the crabs.

Boil for about 15 minutes or just until the crabs are cooked through; drain. Let stand until cool and clean.

Serves 6

The agricultural economy depends upon the research and extension work of the LSU AgCenter. For example, new crop varieties must be developed and tested and made ready for use as old varieties become susceptible to diseases and pests. Crop varieties rarely last more than 10 to 15 years. Without AgCenter researchers and extension specialists developing and testing varieties that will grow in Louisiana, not a single crop would be profitable to produce.

Boiled Crawfish

2 (16-ounce) packages crawfish, crab and
shrimp boil
1 (8-ounce) bottle liquid crab boil
1 (26-ounce) box salt
5 pounds potatoes
3 pounds onions
4 lemons, cut into halves
40 pounds crawfish, rinsed
8 ears of corn, shucked, silk removed

Fill an 84-quart crawfish cauldron about half
full of water. Add the crawfish boil, liquid crab
boil and salt to the water and mix well. Add the
potatoes, onions and lemons. Bring to a boil.
Add the crawfish and cover.

Bring to a boil. Boil for 5 minutes. Turn off
the heat. Add the corn. Let stand, covered, for
20 minutes. Drain and enjoy. Optional
ingredients include 4 or 5 garlic cloves, link
sausage, artichokes and whole mushrooms.

Serves 8

Crawfish Easy

1 pound crawfish or small shrimp
1 1/2 cups rice
1 (14-ounce) can chicken broth
3/4 cup chopped bell pepper
1/2 cup chopped celery
1/2 cup chopped fresh parsley
1 (4-ounce) can mushrooms
1/2 cup (1 stick) butter, melted
5 or 6 green onions, chopped
Seasoned salt to taste
Garlic to taste
Red pepper to taste
Cajun seasoned salt to taste

Spray a rice cooker with nonstick cooking spray.
Add the crawfish, rice, broth, bell pepper,
celery, parsley, undrained mushrooms, butter
and green onions and mix well. Stir in the
seasoned salt, garlic, red pepper and Cajun
seasoned salt and mix well. Cook, covered, for
30 minutes.

Serves 4

*There is no better way to celebrate spring than having a crawfish boil...complete
with onions, potatoes, and corn. Crawfish are a Cajun culinary favorite. These
little crustaceans (the Cajun tall tale claims they were lobsters worn to a frazzle
after following the Acadians from Canadian waters) are eaten boiled like shrimp, in
a bisque, stuffed in green bell peppers, and as étouffée served over rice.*

Crawfish Étouffée

2 teaspoons chicken
bouillon granules
1/4 cup warm water
1/2 cup (1 stick) butter
3 tablespoons flour
1 large onion, chopped
1 small bell pepper, chopped
2 ribs celery, chopped

2 teaspoons minced garlic
1 pound peeled crawfish tails
2 tablespoons tomato paste
1/2 cup water
2 teaspoons Cajun
seasoned salt
6 green onion tops,
finely chopped

Dissolve the bouillon granules in 1/4 cup warm water. Heat the butter in a large saucepan. Stir in the flour. Cook over medium-high heat until light brown, stirring constantly. Stir in the onion, bell pepper, celery and garlic. Add the bouillon and mix well.

Cover and bring to a boil. Boil for 20 minutes. Add the crawfish tails, including fat and juice, tomato paste, 1/2 cup water and Cajun seasoned salt. Cover and bring to a boil. Boil for 20 minutes. Stir in the green onion tops. Cook for 8 minutes, stirring occasionally. Serve over hot cooked rice.

Serves 4

Louisiana leads the nation in the amount of crawfish produced and the LSU AgCenter is the source of nearly all the research and extension work on crawfish. Crawfish are not only delicious, but also extremely high in nutritional value. They are an excellent source of protein, calcium, phosphorus, iron, and the B vitamins. Without a doubt, Louisiana's crawfish are "heads and tails" above crawfish produced in the rest of the country.

South Louisiana Crawfish Fettuccini

2 medium yellow onions, chopped
1 bell pepper, chopped
4 ribs celery, chopped
1 garlic clove, minced
1 cup (2 sticks) margarine
2½ pounds peeled crawfish tails
1 (15-ounce) jar jalapeño processed cheese

2 cups half-and-half
1 bunch green onions, chopped
16 ounces thin noodles, cooked, drained
2 cups shredded Cheddar cheese

Sauté the onions, bell pepper, celery and garlic in the margarine in a saucepan for 2 minutes. Stir in the crawfish tails, processed cheese, half-and-half, green onions and noodles.

Spoon the crawfish mixture into a 9×13-inch baking pan. Sprinkle with the shredded cheese. Bake at 350 degrees for 30 minutes or until brown and bubbly.

Serves 8

Blue Ribbon Crawfish Quiche

1 unbaked (9-inch) pie shell
1 tablespoon flour
1 tablespoon sugar
1/2 cup chopped onion
2 tablespoons butter
8 ounces peeled crawfish tails
Salt and pepper to taste
2 tablespoons chopped
  green onions

2 tablespoons chopped fresh
  parsley
1 cup light cream
4 eggs
4 ounces Swiss cheese,
  shredded

Pierce the side and bottom of the pie shell with a fork. Rub the bottom of the pie shell with a mixture of the flour and sugar. Bake at 350 degrees for 5 minutes.

Sauté the onion in the butter in a skillet until tender. Stir in the crawfish tails. Simmer until the crawfish is cooked through, stirring occasionally. Season with salt and pepper.

Spoon the crawfish mixture into the pie shell. Sprinkle with the green onions and parsley. Whisk the light cream and eggs in a bowl until blended and pour over the crawfish mixture. Sprinkle with the cheese.

Bake at 425 degrees for 15 minutes. Reduce the heat to 300 degrees. Bake for 15 minutes longer. Let stand for 10 minutes before serving. Garnish with additional fresh parsley.

Serves 6

Crawfish Supreme

Crawfish

1/2 cup chopped onion
1/2 cup chopped celery
1/2 cup chopped
fresh parsley
1 cup (2 sticks) margarine

1 pound peeled crawfish
tails, chopped
1 (4-ounce) can
mushrooms, drained
1/2 cup sherry

Juice of 1 lemon
2 pimentos, chopped
1 tablespoon hot sauce
1 teaspoon salt

Cream Sauce and Assembly

1/2 cup flour
1 teaspoon salt
1 teaspoon pepper

2 cups milk
1/2 cup (1 stick) margarine
1/2 cup cracker crumbs

For the crawfish, sauté the onion, celery and parsley in the margarine in a saucepan until tender but not brown. Remove from heat. Stir in the crawfish tails, mushrooms, sherry, lemon juice, pimentos, hot sauce and salt.

For the sauce, mix the flour, salt and pepper in a double boiler. Add the milk gradually, stirring constantly. Cook until thickened, stirring constantly. Remove from heat. Stir in the margarine.

Add the sauce to the crawfish mixture and mix well. Spoon the crawfish mixture into a baking dish. Sprinkle with the cracker crumbs. Bake at 325 degrees for 20 to 25 minutes or until bubbly. Serve over egg noodles or fluffy hot cooked Louisiana rice.

Serves 6

The Breaux Bridge Crawfish Festival began in 1960. The following year, the state legislature named Breaux Bridge the "Crawfish Capital of the World." The festival was held every other year until 1989, when it became an annual event.

Asparagus and Shrimp Oriental

1 (11-ounce) can asparagus
10 medium shrimp, peeled,
deveined
1/4 cup chopped red
bell pepper
1 tablespoon sesame seeds

1 tablespoon sesame oil
2 tablespoons soy sauce
2 tablespoons lemon juice
1/8 teaspoon ginger
1/8 teaspoon Tabasco sauce

Drain and rinse the asparagus. Arrange in a serving dish. Cut each shrimp into halves or thirds. Sauté the shrimp, bell pepper and sesame seeds in the sesame oil in a skillet for 5 minutes or until the shrimp turn pink, stirring constantly. Remove from heat.

Stir the soy sauce, lemon juice, ginger and Tabasco sauce into the shrimp mixture. Spoon the shrimp mixture over the asparagus. Serve immediately. You may substitute fresh steamed asparagus for the canned asparagus.

Serves 4

Cocodrie Baked Shrimp

1/2 cup (1 stick) butter
3 tablespoons bacon drippings
1 cup chopped green onions
1/2 cup chopped celery
4 garlic cloves, chopped
2 (16-ounce) bottles Catalina salad dressing
1 cup crumbled crisp-cooked bacon
1 cup reduced shrimp stock
3 tablespoons Worcestershire sauce
1/2 cup white wine

2 tablespoons Louisiana hot sauce
Juice of 1 lemon
1 tablespoon Creole mustard
4 ounces cream cheese, cubed
1/2 cup shredded Velveeta cheese
2 to 3 pounds medium shrimp, peeled, deveined
1 pound new potatoes, cooked, cut into chunks
Chopped fresh parsley

Heat the butter and bacon drippings in a skillet until the butter melts. Sauté the green onions, celery and garlic in the butter mixture until the green onions and celery are tender. Add the salad dressing, bacon, stock, Worcestershire sauce, wine, hot sauce, lemon juice and Creole mustard and mix well.

Simmer until slightly thickened and reduced, stirring occasionally. Stir in the cream cheese and Velveeta cheese. Cook until the cheeses melt, stirring frequently. Remove from heat.

Arrange the shrimp and new potatoes in a 9×13-inch baking dish. Pour the cheese sauce over the shrimp mixture, turning to coat. Bake at 350 to 400 degrees for 6 to 10 minutes or until the shrimp turn pink. Sprinkle with parsley. You may substitute fish fillets or oysters for the shrimp or use a combination of shrimp, fish and oysters. If substituting fish fillets for the shrimp, use half the sauce. You may substitute bottled clam juice for the shrimp stock.

Serves 8

Herbed Shrimp and Feta Cheese Casserole

8 ounces feta cheese, crumbled
5 1/3 ounces reduced-fat Swiss
cheese, shredded
1 cup evaporated skim milk
1 cup plain nonfat yogurt
2 eggs, beaten
1/3 cup chopped fresh parsley
1 teaspoon basil
1 teaspoon oregano

4 garlic cloves, minced
8 ounces angel hair pasta,
cooked, drained
1 (16-ounce) jar mild chunky
salsa
1 pound medium shrimp,
peeled, deveined
8 ounces part-skim mozzarella
cheese, shredded

Combine the feta cheese, Swiss cheese, evaporated skim milk, yogurt, eggs, parsley, basil, oregano and garlic in a bowl and mix well.

Line the bottom of a 9×13-inch baking dish sprayed with nonstick cooking spray with half the pasta. Spread with the salsa. Top with half the shrimp. Layer with the remaining pasta.

Pour the cheese and egg mixture over the prepared layers. Sprinkle with the mozzarella cheese. Bake at 350 degrees for 30 minutes. Let stand for 10 minutes before serving. Garnish with additional chopped fresh parsley. The equivalent amount of egg substitute may be used as a substitution for the fresh eggs.

Serves 12

Shrimp My Way

1 medium onion, chopped
4 green onions, chopped
4 ribs celery, chopped
1/4 cup (1/2 stick) butter or margarine
1 to 2 pounds shrimp, peeled, deveined
1 (14-ounce) can diced tomatoes with
green chiles
Salt and pepper to taste
Garlic powder to taste

Sauté the onion, green onions and celery in the butter in a large skillet until tender. Stir in the shrimp, undrained tomatoes, salt, pepper and garlic powder.

Simmer for 30 minutes, stirring occasionally. Serve over hot cooked rice.

Serves 4

Sopping Good Shrimp

3 pounds (36 to 40 count) unpeeled shrimp
1 cup (2 sticks) butter
1/4 cup dry white wine
1/4 teaspoon Worcestershire sauce
Garlic powder to taste
Pepper to taste
Salt to taste

Place the shrimp in a 7×11-inch baking dish. Dot with the butter. Pour the wine and Worcestershire sauce over the top. Sprinkle generously with garlic powder and pepper. Season with salt.

Place the baking dish on the middle oven rack. Bake at 375 degrees for 20 minutes, stirring every 5 minutes. The shrimp are ready when the shell at the top of the back separates from the body. The cooking time is determined by the size of the shrimp. Serve the shrimp with hot crusty French bread to "sop up" the sauce.

Serves 6

Boiled Shrimp

¹/₂ cup vegetable oil
¹/₂ cup ketchup (optional)
¹/₄ cup vinegar
1 large onion, cut into quarters
1 lemon, sliced
3 garlic cloves
¹/₂ teaspoon Tabasco sauce
¹/₂ teaspoon pepper
5 pounds unpeeled shrimp, rinsed
1 tablespoon liquid crab boil
³/₄ cup salt

Fill a 6-quart stockpot half full of water. Bring to a boil. Add the oil, ketchup, vinegar, onion, lemon, garlic, Tabasco sauce and pepper and mix well. Bring to a boil. Add the shrimp and liquid crab boil.

Boil for 5 minutes. Remove from heat. Add the salt and stir until dissolved. Let stand, covered, for 30 minutes. Remove the shrimp with a slotted spoon to a serving bowl. Serve immediately with your favorite cocktail sauce or chill before serving. Use the leftover liquid to boil potatoes or corn if desired.

Serves 10

Photograph for this recipe appears on page 29.

Shrimp in Beer

1 medium onion, chopped
1 garlic clove, minced
3 tablespoons vegetable oil
2 pounds unpeeled shrimp, rinsed
1 (12-ounce) can beer
2 tablespoons chopped fresh parsley
1 tablespoon salt
1 bay leaf
¹/₈ teaspoon hot sauce

Sauté the onion and garlic in the oil in a skillet for 2 minutes. Stir in the shrimp, beer, parsley, salt, bay leaf and hot sauce.

Simmer for 12 minutes, stirring occasionally. Remove from heat. Let stand for 1 hour before serving; drain. Serve the shrimp with your favorite dipping sauce and hot crusty French bread.

Serves 4

Shrimp are Louisiana's most valuable fishery product, representing 11 percent of the pounds of marine fisheries' landings, but more than 65 percent of the value.

Shrimp and Crab Linguini

1/4 cup (1/2 stick)
reduced-fat butter
1/2 cup chopped green onions
1/3 cup chopped celery
1/2 cup chopped bell pepper
1 tablespoon flour
1 cup evaporated skim milk
1 (6-ounce) roll garlic cheese,
cut into chunks
4 ounces hot Pepper Jack
cheese, shredded

1 teaspoon parsley flakes
1 teaspoon Creole seasoning
1 pound (30 to 40 count)
deveined peeled cooked
shrimp
1 pound lump crab meat,
drained
8 ounces linguini, cooked,
drained

Heat the butter in a heavy saucepan. Stir in the green onions, celery and bell pepper. Sauté until the vegetables are tender. Stir in the flour. Add the evaporated skim milk gradually, stirring constantly. Cook until thickened, stirring constantly. Stir in the garlic cheese and Pepper Jack cheese.

Cook until the cheese melts, stirring constantly. Stir in the parsley flakes and Creole seasoning. Add the shrimp, crab meat and pasta and mix gently. Simmer for 15 minutes, stirring occasionally.

Serves 8

Shrimp and Crab-Stuffed Peppers

5 large bell peppers
1/2 cup (1 stick) margarine
3 large onions, chopped
2 ribs celery, chopped
1/2 bell pepper, chopped
5 garlic cloves, chopped
Salt and pepper to taste
2 pounds shrimp, peeled, deveined, chopped

2 sleeves saltine crackers, crushed
3 eggs, beaten
1 pound crab meat
3/4 cup bread crumbs
10 teaspoons margarine

Slice 5 bell peppers horizontally into halves. Discard the seeds and membranes. Heat 1/2 cup margarine in a skillet. Stir in the onions, celery, 1/2 bell pepper, garlic, salt and pepper.

Cook until the onions are tender, stirring frequently. Stir in the shrimp. Cook for 3 minutes, stirring frequently. Add the cracker crumbs and eggs, stirring constantly. Remove from heat. Stir in the crab meat.

Spoon the crab meat mixture into the bell pepper halves and arrange in a baking pan. Sprinkle each bell pepper half equally with bread crumbs and dot each with 1 teaspoon of the margarine. Bake at 350 degrees for 45 minutes. Add a small amount of water to the baking pan for more tender bell peppers.

Serves 5

Shrimp Casserole

3 slices bread
1 pound deveined peeled
  shrimp
1 onion, chopped
1 bell pepper, chopped
1/2 cup (1 stick) butter

1 (14-ounce) can diced
  tomatoes with green chiles
1 (10-ounce) can cream of
  mushroom soup
3 cups cooked rice

Soak the bread in enough water to cover in a bowl; drain. Sauté the shrimp, onion and bell pepper in the butter in a skillet. Stir in the undrained tomatoes and soup. Cook for 5 minutes, stirring frequently. Add the bread and rice and mix well.

Spoon the shrimp mixture into a 2-quart baking dish. Bake at 350 degrees for 30 minutes. Serve with a mixed green salad and hot crusty French bread.

Serves 6

A tree is an agricultural crop, too. Though most lumber leaves the state to be made into furniture, cabinetry and other wood structures, the Louisiana Forest Products Laboratory, part of the LSU AgCenter, provides an array of workshops and educational programs to help expand and develop the secondary wood products industry here so that economic activity will remain at home.

Shrimp Fettuccini

2 pounds shrimp
2 quarts water
1/2 cup (1 stick) butter
1/2 cup minced onion
1 1/2 tablespoons flour
3 cups cream

2 cups chicken broth
1/2 cup white wine
1/2 cup grated Parmesan cheese
2 tablespoons Creole mustard
16 ounces fettuccini

Peel the shrimp, reserving the shells. Bring the water and reserved shrimp shells to a boil in a stockpot. Boil for 25 minutes. Strain, discarding the shells and reserving the stock.

Heat the butter in a heavy saucepan. Sauté the onion in the butter until tender. Sprinkle the flour over the onion mixture and mix well. Add 2 cups of the reserved shrimp stock, cream, broth and wine gradually and mix well.

Cook over low heat for 10 minutes or until thickened, stirring constantly. Stir in the cheese and Creole mustard. Add the shrimp and mix well.

Cook over low heat for 10 minutes, stirring frequently. Cook the pasta in the remaining shrimp stock using package directions; drain. Spoon the shrimp mixture over the fettuccini on a serving platter.

Serves 8

Shrimp Étouffée

2 large onions, finely chopped
1 large bell pepper, finely chopped
1/2 cup (1 stick) butter
2 pounds shrimp, peeled, deveined
Chopped green onions to taste
Salt and pepper to taste

Sauté the onions and bell pepper in the butter in a skillet until tender. Stir in the shrimp. Cook over medium heat for 10 to 15 minutes or until the shrimp turn pink, stirring frequently and adding small amounts of water as needed. Add green onions and mix well.

Cook for 5 minutes longer, stirring occasionally. Season with salt and pepper. Serve over hot cooked rice.

Serves 6

Barbecued Shrimp Kabobs

1 (10-ounce) bottle Heinz 57 Steak Sauce
1/2 cup (1 stick) butter or margarine
1/3 cup lemon juice
1/4 cup Worcestershire sauce
1 pound deveined peeled medium shrimp
2 onions, cut into chunks
2 bell peppers, cut into chunks
2 yellow squash, thickly sliced
10 cherry tomatoes

Combine the steak sauce, butter, lemon juice and Worcestershire sauce in a saucepan. Simmer over medium heat until heated through, stirring occasionally.

Thread the shrimp, onions, bell peppers, squash and cherry tomatoes alternately on skewers. Brush with the steak sauce mixture.

Grill the kabobs over hot coals until the shrimp turn pink and the vegetables are the desired degree of crispness, basting frequently with the remaining steak sauce mixture. You may substitute your favorite fresh vegetables for the onions, bell peppers, squash and cherry tomatoes if desired.

Serves 6

Shrimp Pie

1/4 cup (1/2 stick) margarine
1 cup chopped onion
1 cup chopped mushrooms
1/2 cup chopped bell pepper
1/2 teaspoon grated lemon zest
1 pound shrimp, peeled, deveined
1/2 cup half-and-half
1/2 cup grated Parmesan cheese

1 tablespoon paprika
1/2 teaspoon black pepper
1/8 teaspoon red pepper
1/3 cup flour
1 (baked) 9-inch pie shell
1/3 cup shredded mozzarella cheese
1/3 cup shredded Cheddar cheese

Heat the margarine in a nonstick saucepan. Add the onion, mushrooms, bell pepper and lemon zest. Cook over medium heat until the vegetables are tender, stirring frequently. Stir in the shrimp, half-and-half and Parmesan cheese.

Cook over medium heat for 3 minutes or until the shrimp turn pink, stirring frequently. Stir in the paprika, black pepper and red pepper. Combine the flour with a small amount of cold water in a bowl and stir until the consistency of a thick paste. Add the flour mixture to the shrimp mixture, stirring constantly. Cook until thickened, stirring constantly.

Spoon the shrimp mixture into the pie shell. Sprinkle with the mozzarella cheese and Cheddar cheese. Bake at 350 degrees for 45 minutes.

Serves 8

Shrimp with Rice

1/2 cup (1 stick) butter
1 onion, finely chopped
1 cup sliced canned
  mushrooms
1 green bell pepper, chopped
1 1/4 cups rice
1 teaspoon salt
1/2 teaspoon pepper
1/4 teaspoon nutmeg

1 cup dry white wine
3 cups hot water
2 tablespoons chopped fresh
  parsley
1/2 bay leaf
1/4 teaspoon thyme
2 pounds deveined peeled
  shrimp

Heat the butter in a Dutch oven. Add the onion, mushrooms, bell pepper, rice, salt, pepper and nutmeg and mix well. Cook until the rice is golden brown, stirring constantly. Stir in the wine.

Simmer for 5 minutes, stirring occasionally. Add the hot water, parsley, bay leaf and thyme and mix well. Cook, covered, for 10 minutes, stirring occasionally. Stir in the shrimp. Simmer for 5 to 10 minutes or until the shrimp turn pink, stirring occasionally. Let stand, covered, for 5 to 10 minutes. Discard the bay leaf and serve.

Serves 6

Seafood au Gratin

1 cup chopped onion
1/2 small bell pepper,
  finely chopped
1 rib celery, finely chopped
3 tablespoons chopped
  fresh parsley
2 garlic cloves, minced
1/4 cup (1/2 stick) butter
1/4 cup flour
2 cups milk

2 egg yolks, beaten
1 teaspoon salt
1/2 teaspoon red pepper
1/4 teaspoon black pepper
12 ounces crab meat
12 ounces deveined peeled
  shrimp
3/4 cup shredded Cheddar
  cheese

Sauté the onion, bell pepper, celery, parsley and garlic in the butter in a heavy saucepan until the vegetables are tender but not brown. Stir in the flour. Cook until bubbly, stirring constantly. Add the milk gradually, stirring constantly. Cook until thickened, stirring constantly.

Stir a small amount of the hot mixture into the egg yolks. Stir the egg yolks, salt, red pepper and black pepper into the thickened mixture. Cook over medium heat for 5 minutes, stirring constantly. Fold in the crab meat and shrimp.

Spoon the crab meat mixture into a baking dish. Sprinkle with the cheese. Bake at 350 degrees until brown and bubbly.

Serves 6

Au gratin refers to the crusty cheese topping created on top of a sauce or casserole after it is removed from the oven or broiler. The most famous of all au gratins in Louisiana is the Jumbo Lump Crab Meat au Gratin. Shrimp and crawfish may be added to further enhance the dish.

Seafood Potato Casserole

5 pounds potatoes, peeled
Salt to taste
Liquid crab boil (optional)
2 tablespoons canola oil
2 onions, chopped
1/2 bell pepper, chopped
1/2 rib celery, finely chopped
Water or shrimp stock
1 (6-ounce) can crab meat

1 pound deveined peeled
fresh or frozen small shrimp
1 pound peeled fresh or
frozen crawfish
1 cup coarsely chopped fish
fillet (optional)
Cajun seasoned salt to taste
Pepper to taste
Garlic powder to taste

1/4 cup grated Parmesan
cheese
8 ounces each Cheddar and
mozzarella cheese,
shredded
1/2 bunch green onions,
thinly sliced
2 tablespoons butter or
margarine (optional)

Cut each potato into 6 or 8 chunks. Combine the potatoes, salt and a few drops of liquid crab boil with enough water to cover in a saucepan. Bring to a boil. Boil until the potatoes are tender; drain. Cut the potatoes into smaller chunks and place in a bowl. Cover to keep warm.

Heat the canola oil in a saucepan over medium heat. Add the onions, bell pepper and celery to the hot oil. Cook, covered, until the vegetables are very tender or almost mushy, adding water or shrimp stock if needed to prevent the vegetables from sticking or burning. Stir in the crab meat, shrimp, crawfish, fish and a few drops of liquid crab boil. Thaw frozen seafood before adding. Cook, covered, over medium heat until the shrimp turn pink and the fish flakes easily, stirring frequently. Season with the next 3 ingredients. Simmer, covered, until serving time adding water or shrimp stock if the mixture begins to stick.

Drain the seafood mixture, reserving the liquid. Add the seafood mixture to the potatoes and mix gently, adding the reserved liquid as desired if the mixture is too dry. Stir in the Parmesan cheese. Stir in half the Cheddar cheese and half the mozzarella cheese. Fold in the scallions. Drop the seafood and potato mixture by spoonfuls into 3 or 4 greased 9×13-inch baking pans; do not smooth. Dot with the butter and sprinkle with the remaining cheese. Bake at 350 degrees for 30 minutes or until bubbly and light brown.

Serves 20

If you buy fresh or frozen shrimp in the shell, you can prepare a good, inexpensive Shrimp Stock. Peel and devein the shrimp, reserving the heads and shells. Chill the shrimp for later use. Mix the reserved heads and shells with enough water to cover in a saucepan. Stir in one or two drops of liquid crab boil. Bring to a boil over high heat; reduce the heat. Simmer for 5 minutes. Strain the stock, discarding the solids.

Rice Cooker Seafood Jambalaya

1 pound deveined peeled
shrimp
1 pint oysters
2 cups crab meat, drained
1 1/2 cups rice
(rice cooker cups)
1 medium bell pepper,
chopped

1 medium onion, chopped
1 (10-ounce) can beef broth
1 (6-ounce) can mushrooms,
drained
1/2 cup (1 stick) butter or
margarine, melted
Salt and pepper to taste

Combine the shrimp, undrained oysters, crab meat, rice, bell pepper, onion, broth, mushrooms, butter, salt and pepper in a rice cooker and mix well. Cook using manufacturer's directions.

Serves 6

Have a question about nutrition or special diets, food preparation, ingredient substitutions, or food safety? Check our website at www.agctr.lsu.edu/eatsmart/fpg.htm.

Louisiana Baked Oysters

2 quarts oysters, drained
1/4 cup chopped fresh
parsley
1/2 cup chopped green onions
2 tablespoons lemon juice
1 tablespoon Worcestershire
sauce

1/2 cup (1 stick) butter or
margarine, melted
Salt and pepper to taste
Tabasco sauce to taste
2 cups cracker crumbs
Paprika to taste
3/4 cup milk

Layer the oysters, parsley, green onions, lemon juice, Worcestershire sauce, butter, salt, pepper, Tabasco sauce and cracker crumbs 1/2 at a time in a greased 8×12-inch baking dish. Sprinkle with paprika.

Poke several holes in the top. Pour the milk into the holes, being careful not to moisten the crumb topping all over. Bake at 375 degrees for 10 minutes or until firm.

Serves 12

Cocktail Sauce

3 medium onions, minced
1 (14-ounce) bottle
ketchup
2 cups mayonnaise
2 tablespoons olive oil

1 teaspoon dry mustard
1 teaspoon horseradish
1 teaspoon Worcestershire
sauce

Combine the onions, ketchup, mayonnaise, olive oil, dry mustard, horseradish and Worcestershire sauce in a bowl and mix well. Serve as a dipping sauce with seafood or spoon over a mixed green salad.

Makes 16 (1/4-cup) servings

Oysters à la LuLa

1 pint oysters
Salt and pepper to taste
1/4 to 1/3 cup flour
2 tablespoons olive oil

2 garlic cloves, minced
1/4 cup chopped shallots
1/2 cup dry white wine
4 slices bread, toasted

Drain the oysters and pat dry with paper towels. Sprinkle the oysters with salt and pepper and coat lightly with the flour. Heat the olive oil in an 8-inch nonstick skillet over medium heat. Add the oysters and garlic.

Cook for 3 minutes or until the edges of the oysters curl and are light brown, stirring occasionally. Remove the oysters with a slotted spoon to a heated platter, reserving the pan drippings.

Add the shallots to the reserved pan drippings and mix well. Cook for about 2 minutes, stirring frequently. Stir in the wine. Return the oysters to the skillet. Simmer for 2 to 3 minutes or until heated through, stirring frequently. Spoon the oysters and sauce over the toasted bread. You may substitute 1/2 teaspoon garlic powder for the fresh garlic and green onions for the shallots.

Serves 4

A partnership between AgCenter scientists and entrepreneurs in Louisiana's oyster industry has resulted in a revival of the Gulf Coast raw oyster. Louisiana had been a key supplier of this product, but its marketing was threatened by fears that a deadly microorganism might be lurking inside the shell. Through a heating process similar to the pasteurization of raw milk, the offending microorganism was killed without hurting the texture or flavor of the raw oyster.

Alligator Sauce Piquant

1 cup olive oil
3 cups flour
5 cups chopped onions
2 cups chopped green onions
1 cup chopped bell pepper
1/2 cup chopped celery
2 cups chopped tomatoes
8 cups cold water
6 cups tomato sauce
2 cups dry white wine

2 tablespoons finely
  chopped garlic
2 tablespoons Worcestershire
  sauce
Juice of 1 lemon
Salt to taste
Louisiana hot sauce or
  cayenne pepper to taste
5 pounds alligator meat,
  trimmed, cubed

Heat the olive oil in a cast-iron skillet. Add the flour gradually, stirring constantly. Cook until dark brown in color and of a roux consistency, stirring constantly. Stir in the onions, green onions, bell pepper and celery.

Cook, covered, until the onions are tender, stirring occasionally. Stir in the tomatoes. Cook for 10 minutes, stirring frequently. Add the cold water and mix well. Cook until thickened, stirring constantly. Stir in the tomato sauce, wine, garlic, Worcestershire sauce, lemon juice, salt and hot sauce. Add the alligator meat and enough water to cover by 2 inches and mix well.

Bring to a boil, stirring frequently. Reduce the heat to low. Cook, covered, for 3 to 4 hours or until the alligator meat is tender, stirring occasionally. Serve over hot cooked rice or spaghetti with grated Parmesan cheese.

Serves 20

Alligators and fashion may bring different images to mind, but the combination offers potential for Louisiana's economy. The AgCenter's School of Human Ecology has initiated research to explore ways to increase domestic demand for finished products made with American alligator leather, while the Department of Food Science is researching nutritional aspects of alligator.

Serving Louisiana
Vegetables
&
Side Dishes

Vegetables & Side Dishes

Asparagus Casserole

1 (10-ounce) package frozen
green peas
3 or 4 carrots, sliced
2 (10-ounce) cans asparagus,
drained

1 (10-ounce) can cream of
mushroom soup
1 (10-ounce) can cream of
celery soup

Cook the peas using package directions until tender-crisp; drain. Combine the carrots with enough water to cover in a saucepan. Bring to a boil; reduce the heat. Cook until the carrots are tender-crisp; drain.

Arrange 1 can of the asparagus over the bottom of a buttered baking dish. Layer with the peas, carrots and remaining can of asparagus.

Mix the soups in a bowl. Spread the soup mixture over the prepared layers. Bake, covered, at 350 degrees for 30 minutes or until bubbly.

Serves 6

Look for asparagus with a fresh green color. The tips should be compact, tightly closed, and dark green or purplish in color. Avoid fibrous stalks with spread-open tips. Most people instinctively reach for the young, slender stalks, but the thicker, more mature stalks are considered the connoisseur's delight…more tender and flavorful. Asparagus contains vitamin A, folic acid, and calcium.

Red Beans and Rice

1 pound dried red beans or pinto beans
1 onion, chopped
1 bell pepper, chopped
2 garlic cloves, minced
8 to 10 cups water
1 pound hot reduced-fat sausage
1 pound mild reduced-fat sausage
2 (16-ounce) cans tomatoes
1 teaspoon chili powder
Lite salt to taste

Sort and rinse the beans. Mix the beans, onion, bell pepper, garlic and water in a stockpot. Bring to a boil; reduce the heat. Cook, covered, until the beans are tender, stirring occasionally.

Sauté the sausage in a skillet; drain. Add the sausage, undrained tomatoes, chili powder and lite salt to the beans and mix well. Simmer for 20 to 30 minutes, stirring occasionally. Serve over hot cooked rice.

Serves 10

Ginger Beets

6 cups small beets
2 cups sugar
2 tablespoons flour
4 teaspoons ginger
1 cup apple cider
1 cup raisins
1 tablespoon butter or margarine
1/2 teaspoon salt (optional)

Combine the beets with enough water to cover in a saucepan. Cook until tender. Drain, reserving 1 cup of the liquid. Peel the beets.

Combine the sugar, flour and ginger in a saucepan and mix well. Stir in the apple cider and reserved beet liquid. Bring to a boil over high heat, stirring constantly. Boil until thickened, stirring constantly. Stir in the raisins, butter, salt and beets. Cook just until heated through, stirring occasionally. You may substitute a mixture of 1/2 cup apple juice and 1/2 cup vinegar for the apple cider.

Serves 8

Beans have long been known for their low cost and high nutritional value. A pound of cooked dried beans will make about nine servings, compared to four servings per pound of meat, poultry, or fish. Serving beans is a good way to stretch the food dollar. Beans are available in several forms: dried, canned, frozen, and for a limited period during the growing season, fresh. Louisiana is famous for red beans and rice.

Broccoli and Rice Casserole

1 large onion, chopped
1 tablespoon margarine
2 (10-ounce) packages frozen
chopped broccoli, thawed,
drained
1 (10-ounce) can cream of
mushroom soup
1/2 soup can water

1 (8-ounce) jar processed
cheese
1/4 cup milk
2 cups cooked rice
Salt and pepper to taste
1 (6-ounce) can
French-fried onions

Sauté the chopped onion in the margarine in a saucepan until tender. Stir in the broccoli. Add the soup, water, cheese and milk and mix well. Cook until the cheese melts, stirring frequently. Stir in the rice. Season with salt and pepper.

Spoon the broccoli mixture into a lightly greased 11×13-inch baking dish. Bake at 350 degrees for 30 minutes. Sprinkle with the French-fried onions. Bake just until the onions brown. Serve immediately. You may freeze for future use. Sprinkle with the French-fried onions just before baking and extend the baking time to 45 minutes.

Serves 6

Broccoli is such a versatile vegetable. Its color and taste complement many main dishes, plus it is undeniably packed with nutrition. Broccoli is an excellent source of vitamin C and vitamin A, two of our antioxidant vitamins which researchers believe offer protection against heart disease, stroke, and certain types of cancer. And it is low in calories.

Glazed Carrots

3 (12-ounce) packages small whole carrots,
peeled
1/4 cup honey
1/4 cup (1/2 stick) butter or margarine, melted
1/4 cup packed brown sugar

Combine the carrots with enough water to
cover in a saucepan. Bring to a boil. Boil for 10
minutes or until tender-crisp. Drain, reserving
1/4 cup of the liquid.

Combine the reserved liquid, honey, butter
and brown sugar in a bowl and mix well. Pour
the butter mixture over the carrots and stir until
coated. Cook over low heat until heated
through, stirring occasionally.

Serves 6

Corn Pudding

2 (17-ounce) cans cream-style corn
1 (12-ounce) can whole kernel corn
1/4 to 1/2 cup sugar
1/2 cup milk
1/2 cup (1 stick) butter, melted
5 eggs, lightly beaten
1/4 cup cornstarch
1 teaspoon minced onion
1/2 teaspoon dry mustard

Combine the corn, sugar, milk, butter, eggs,
cornstarch, onion and dry mustard in a bowl
and mix well. Spoon the corn mixture into a
3-quart baking dish. Bake at 400 degrees for
1 hour, stirring halfway through the baking
process.

Serves 8

*The good ol' carrot is a vegetable that can be found on the market all year
with no distinct peak time. Carrots are packed with vitamin A. Vitamin A is
important for good eyesight, especially at night, healthy skin, and
germ-resistant lining of the nose, throat, and lungs. Raw carrots, eaten as
a snack, can actually help clean the teeth!*

Louisiana Corn Creole

1 medium green bell pepper,
chopped
1 small onion, chopped
3 tablespoons butter
1 pound ground beef
1 (17-ounce) can
cream-style corn

2 cups milk
1 egg, beaten
1/3 to 1/2 cup cornmeal
1 teaspoon salt
1 teaspoon pepper
1/2 cup bread crumbs
2 tablespoons butter

Sauté the bell pepper and onion in 3 tablespoons butter in a skillet. Add the ground beef. Cook until the ground beef is light brown and crumbly; drain. Stir in the corn, milk and egg.

Cook for 5 minutes, stirring frequently. Add the cornmeal, salt and pepper and mix well. Cook until of the consistency of mush, stirring frequently. Spoon the corn mixture into a greased baking dish. Sprinkle with the bread crumbs and dot with 2 tablespoons butter. Bake at 350 degrees for 45 minutes or until firm and brown. Serve with cranberry sauce.

This recipe, used in the Extension Service Food & Nutrition Program, is a one-dish meal, providing all the essential food nutrients. It is economical in that any leftover meat such as link sausage, hot dogs, tuna, crab meat, chicken or shrimp may be substituted for the ground beef.

Serves 6

A summertime favorite, whether on the cob or cream-style, fresh sweet corn is at its sweetest harvested just as the kernels mature. Select corn with fresh, succulent husks and a good green color. The silk ends should be free of decay or worm injury. Ears should be well covered with plump, tender kernels. Freezing the corn as soon as it is ready to eat can capture the sweetness.

Eggplant with Shrimp

5 slices white sandwich
bread, torn
1 cup milk
2 eggplant, peeled, chopped
Salt to taste
2 slices bacon
2 medium onions, chopped
1 medium bell pepper,
chopped

$^1/_2$ cup chopped celery
4 garlic cloves, minced
2 pounds shrimp, peeled,
deveined
$^1/_2$ cup grated Romano cheese
$^1/_4$ cup minced fresh parsley
2 eggs, beaten
1 teaspoon lemon juice
Pepper to taste

Soak the bread in the milk until soft. Combine the eggplant and salt with enough water to cover in a bowl. Let stand for 30 minutes; drain. Combine the eggplant with just enough water to cover in a saucepan. Cook until tender; drain.

Fry the bacon in a skillet until crisp. Drain, reserving the bacon drippings. Crumble the bacon. Sauté the onions, bell pepper, celery and garlic in the reserved bacon drippings. Stir in the shrimp. Cook until the shrimp turn pink.

Squeeze the excess milk from the bread. Add the bread, eggplant, bacon, $^1/_4$ cup of the cheese, parsley, eggs, lemon juice and pepper to the shrimp mixture. Spoon the eggplant mixture into a buttered baking dish. Sprinkle with the remaining $^1/_4$ cup cheese. Bake at 350 degrees for 45 minutes.

Serves 6

Stuffed Mirliton

4 mirliton	1/2 medium bell pepper,	1 1/2 cups seasoned bread
Salt to taste	chopped	crumbs
1 pound ground beef	Pepper to taste	1 tablespoon butter or
1 medium onion, chopped	Cajun seasoned salt to taste	margarine

Boil the mirliton gently in lightly salted water in a saucepan until fork tender; when testing be careful not to tear the skin. Do not overcook as they will be too soft to hollow out and stuff. Let stand until cool. Cut the mirliton lengthwise into halves; discard the seeds. Scoop the pulp into a bowl, leaving a 1/4-inch shell. Chop the pulp into small pieces.

Brown the ground beef in a skillet, stirring until crumbly; drain. Stir in the onion, bell pepper, salt, pepper and Cajun seasoned salt. Simmer, covered, until the vegetables are tender, stirring frequently. Stir in the mirliton pulp.

Cook over medium-high heat until the pulp is mushy and the liquid has been absorbed, stirring frequently. Stir in 1 cup of the bread crumbs. The mixture should be similar to the consistency of a dry paste, which sticks to the spoon. If too moist, add more bread crumbs. Taste and adjust seasonings.

Spoon the ground beef mixture into the shells. Sprinkle each with some of the remaining bread crumbs and dot with the butter. Arrange the shells on a lightly greased baking sheet or in a shallow baking pan. Bake at 350 degrees until golden brown.

For Seafood-Stuffed Mirliton, substitute 1 pound of cooked seafood for the ground beef. Sauté the onion and bell pepper first and then add the cooked seafood. Season to taste. Simmer, covered, to allow the flavors to blend.

Serves 8

In Louisiana, mirliton (MER-lee-tawn) are sometimes called "alligator pears." This is because mirliton are shaped like pears and when held sideways resemble an alligator's head. Mirliton are a variety of summer squash also known as chayotes or vegetable pears. They are mild tasting, like zucchini; however, denser texture requires more cooking time than zucchini.

Okra Croquettes

3 tablespoons cornmeal
1 tablespoon flour
1/2 teaspoon pepper
1/2 teaspoon salt
1 egg, beaten
2 cups finely chopped okra pods
1/2 cup chopped onion
Vegetable oil for deep-frying

Combine the cornmeal, flour, pepper and salt in a bowl and mix well. Stir in the egg. Add the okra and onion and mix well. This mixture has an unusual consistency, so you must press the mixture by tablespoonfuls on the side of the bowl to form the croquettes.

Fry the croquettes in hot oil in a deep-fryer until brown on both sides and the desired degree of crispness, turning once. Drain on paper towels. Serve immediately.

Serves 4

Potatoes Supreme

1 (32-ounce) package frozen hash brown potatoes
2 cups sour cream
1 (10-ounce) can cream of chicken soup
10 ounces Cheddar cheese, shredded
1/2 cup chopped onion
1/2 cup (1 stick) butter, melted
1 teaspoon salt
1/2 teaspoon pepper
2 cups cornflakes, crushed
1/4 cup (1/2 stick) butter, melted

Combine the potatoes, sour cream, soup, cheese, onion, 1/2 cup butter, salt and pepper in a bowl and mix well. Spoon the potato mixture into a buttered 9×13-inch baking dish.

Bake at 350 degrees for 30 minutes. Sprinkle with a mixture of the cornflakes and 1/4 cup butter. Bake for 30 minutes longer. For variety, add 1 1/2 cups chopped green onions and/or 1/2 cup chopped bell pepper.

Serves 12

Okra is a favorite summer vegetable for Louisianians. It is served in a variety of ways…fried, boiled, stewed with tomatoes, and as a thickening agent for soups and gumbos. But pickled okra ranks second to fried okra as a Louisiana favorite. For pickles, harvest when the okra is young and tender. It is best to select short pods for pint and half-pint jars. It is important not to cut into the okra pod because that will result in it getting slimy.

Sweet Potato Casserole with Pecan Topping

6 to 8 medium sweet potatoes
1/2 cup skim milk
1/3 cup sugar, or to taste
1/4 cup applesauce
2 egg whites, beaten
1/4 teaspoon lite salt
Cinnamon to taste

Butter sprinkles to taste
2/3 cup packed brown sugar
1/4 cup flour
1 tablespoon butter sprinkles
1/2 cup chopped pecans
1/4 cup Grape-Nuts cereal

Combine the sweet potatoes with enough water to cover in a saucepan. Bring to a boil. Boil until tender; drain. Peel the sweet potatoes and place in a bowl. Mash until smooth. Add the skim milk, sugar, applesauce, egg whites, lite salt, cinnamon and butter sprinkles to taste and mix well; the mixture should be the consistency of a custard.

Spoon the sweet potato mixture into a 9×11-inch baking dish sprayed with nonstick cooking spray. Mix the brown sugar, flour and 1 tablespoon butter sprinkles in a bowl. Stir in the pecans and cereal. Spoon the brown sugar mixture over the top. Bake at 350 degrees for 1 hour. You may substitute an equivalent amount of artificial sweetener for the sugar.

Serves 10

The LSU AgCenter Sweet Potato Research Station near Chase, Louisiana, is the only sweet potato research station in the country. The LSU AgCenter released a high-yielding variety named "Beauregard." This sweet potato has a sweet, rich flavor when cured and is disease resistant.

Southern-Style Yellow Squash

6 small to medium yellow squash,
cut into 1/2-inch slices
Salt and pepper to taste
1/2 medium onion, sliced, separated into rings
2 tablespoons butter or margarine

Combine the squash, salt and pepper with enough water to cover in a saucepan. Add the onion. Bring to a boil; reduce the heat.

Simmer, covered, for 5 minutes or until the squash is fork tender; do not overcook. Drain half the liquid. Spoon the squash and remaining liquid into a serving bowl. Top with the butter. You may add 8 ounces deveined peeled shrimp before adding the onions for a delicious alternative.

Serves 6

Squash Casserole

2 quarts yellow squash, steamed
1 onion, thinly sliced
1/2 bell pepper, chopped
3 slices bacon, chopped
Sliced jalapeño chile to taste
Salt and pepper to taste
1 (8-ounce) jar processed cheese
1 cup cracker crumbs or seasoned bread crumbs

Spoon half the squash into a baking dish. Sauté the onion, bell pepper and bacon in a skillet until the bacon is crisp; drain. Stir in the jalapeño chile. Spoon the bacon mixture over the squash. Sprinkle with salt and pepper.

Heat the cheese in a saucepan until melted, stirring frequently. Spread half the cheese over the prepared layers and layer with the remaining squash. Spread the remaining cheese over the top and sprinkle with the cracker crumbs. Bake at 350 degrees for 30 to 40 minutes or until brown and bubbly.

Serves 8

You just can't beat fresh summer squash! Squash is very low in calories and is a good source of vitamin C. However, since you lose about half of the nutrients during the cooking process, consume it raw as often as possible. When selecting yellow squash, pick those with fresh looking skin that are free of blemishes. It should be firm and fairly heavy for its size. The smaller, lighter colored squash have the best eating quality.

Fried Green Tomatoes

3 firm green tomatoes
1/2 cup evaporated milk
1/2 cup water
1 egg
1/4 teaspoon salt
1/4 teaspoon pepper
1 cup flour
1 cup seasoned bread crumbs
6 cups vegetable oil
2 garlic cloves, minced
1/4 cup chopped fresh parsley

Cut each green tomato into four 1/2-inch slices. Whisk the evaporated milk, water, egg, salt and pepper in a bowl until blended. Coat the sliced green tomatoes with the flour, dip in the egg mixture and then coat with the bread crumbs.

Fry the breaded tomato slices in the hot oil in a deep-fryer for 5 minutes or until they rise to the top; drain. Sprinkle with the garlic and parsley. Serve immediately.

Makes 12 slices

Zucchini Rice Bake

3 cups sliced zucchini
1 cup sliced mushrooms (optional)
3/4 cup chopped onion
1 garlic clove, chopped
1 tablespoon vegetable oil
3 cups cooked rice
2 cups shredded reduced-fat mozzarella cheese
1 cup chopped cooked chicken or shrimp
2 eggs, beaten
1 tablespoon prepared mustard
1/2 teaspoon pepper
1/2 teaspoon oregano

Sauté the zucchini, mushrooms, onion and garlic in the oil in a skillet for 10 minutes. Stir in the rice, cheese, chicken, eggs, prepared mustard, pepper and oregano. Spoon the zucchini mixture into a 2-quart baking dish. Bake at 350 degrees for 20 minutes.

Serves 6

The tomato is actually a fruit, therefore, it should be treated like one, which means not storing it in the refrigerator. Tomatoes at room temperature have the best flavor. If they must be refrigerated after fully ripening, allow them to come to room temperature before serving. Tomatoes not only pack a lot of vitamin A and C, but they are rich in lycopene, the antioxidant which can help protect against colon, lung, and prostate cancer. Lycopene gives tomatoes their red pigment.

Festive Holiday Yams

8 to 10 small yams
2 cups fresh cranberries
1/4 to 1/3 cup walnuts, chopped
1 cup orange marmalade
1/2 cup packed brown sugar
1/4 cup (1/2 stick) butter

1/4 cup honey
1 tablespoon lemon juice
1 tablespoon orange juice
    concentrate
1/2 teaspoon cinnamon
1/4 cup brandy

Combine the yams with enough water to cover in a saucepan. Bring to a boil. Boil just until tender; drain. Let stand until cool. Peel and cut lengthwise into halves. Arrange the halves overlapping in a 9×13-inch baking dish. Sprinkle with the cranberries and walnuts.

Combine the orange marmalade, brown sugar, butter, honey, lemon juice, orange juice concentrate and cinnamon in a saucepan. Cook until the butter melts and the mixture is of a syrupy consistency, stirring frequently. Remove from heat. Stir in the brandy. Pour over the prepared layers. Bake at 350 degrees for 30 minutes or until bubbly. May be frozen before baking for future use. To serve, thaw and bake.

Serves 10

Photograph for this recipe appears on page 133.

When choosing fresh citrus fruit, lift, don't squeeze. The heavier the fruit is for its size, the juicier it will be. The fruit should be firm. The skin color is not always a good guide when selecting quality fruit, as a green tinge on the skin is quite normal. Citrus fruit, in general, will keep at room temperature for a week to ten days.

Vegetable Batter

1 cup flour
2 tablespoons yellow cornmeal
2 teaspoons salt
2 teaspoons pepper
1 teaspoon sugar
1 cup ice water
2 tablespoons vegetable oil
1 egg, beaten

Combine the flour, cornmeal, salt, pepper and sugar in a bowl and mix well. Stir in the ice water, oil and egg. Dip sliced eggplant, yellow squash, zucchini, green tomatoes or shrimp into the batter and deep-fry until golden brown. You may store the batter, covered, in the refrigerator for several days.

Makes 1 cup

Hot Sherried Fruit

1 (15-ounce) can sliced pineapple, drained
1 (16-ounce) can peach halves, drained
1 (15-ounce) can pear halves, drained
1 (17-ounce) can apricot halves,
drained (optional)
1 (14-ounce) jar apple rings, drained
1 cup sherry
1/2 cup (1 stick) butter
1/2 cup packed light brown sugar
2 tablespoons (heaping) flour

Layer the pineapple, peaches, pears and apricots in a 2-quart baking dish. Top with the apple rings. Combine the sherry, butter, brown sugar and flour in a double boiler. Cook until thickened and smooth, stirring frequently. Pour over the prepared layers.

Chill, covered, for 8 to 10 hours or for up to several days. Bake at 350 degrees for 30 minutes or until bubbly. You may double the recipe for a larger crowd.

Serves 12

Cooking oils, shortenings, and margarine often say "no cholesterol" in bold print...as if it is something unique to that brand. Actually, no vegetable oil contains cholesterol. Cooking oils come from plants, primarily seeds or nuts, and only animal foods contain cholesterol. This label is a marketing gimmick and may be misleading.

Garlic Grits

1/2 cup grits
2 cups water
1/2 teaspoon salt
4 ounces Velveeta cheese, shredded, or to taste
1/2 (6-ounce) roll garlic cheese
1/4 cup (1/2 stick) margarine
1 tablespoon Worcestershire sauce
Garlic powder to taste
Paprika to taste

Cook the grits with the water and salt using package directions. Add the Velveeta cheese, garlic cheese, margarine and Worcestershire sauce to the grits and stir until the cheese melts. Taste and season with garlic powder for a stronger garlic flavor if desired.

Spoon the grits mixture into a 1-quart baking dish. Sprinkle with paprika. Bake at 350 degrees for 15 to 20 minutes or until bubbly.

Serves 4

Crawfish Corn Bread Dressing

2 (8-ounce) packages corn bread mix
1 pound crawfish tails, peeled, chopped
3/4 cup chopped onion
3/4 cup Cajun dressing mix
1/2 cup (1 stick) butter
2 cups reduced-sodium chicken broth
Cajun seasoned salt to taste

Prepare and bake the corn bread using package directions. Let stand until cool and finely crumble. Sauté the crawfish tails, onion and dressing mix in the butter in a skillet for 10 minutes.

Combine the corn bread, crawfish mixture, broth and Cajun seasoned salt in a bowl and mix well. Spoon the corn bread mixture into a 2 1/2- to 3-quart baking dish. Bake at 350 degrees for 15 to 20 minutes or until brown and bubbly.

Serves 6

A baked casserole of cheese grits is one of the most popular side dishes in Louisiana. Some Louisiana cooks add the egg yolks, and then fold in the beaten egg whites at the last minute to make a real soufflé.

Mexican Corn Bread Casserole

1 pound ground chuck
1/2 cup chopped onion
1 envelope taco seasoning mix
2 tablespoons butter
1 (15-ounce) can cream-style corn
1 (8-ounce) package corn muffin mix
1/3 cup milk
1 egg
2 cups shredded Cheddar cheese

Brown the ground chuck with the onion in a skillet, stirring until the ground beef is crumbly and adding the seasoning mix just before the ground chuck is cooked through; drain.

Heat the butter in a large cast-iron skillet until melted. Combine the corn, muffin mix, milk and egg in a bowl and mix well. Pour 1/2 to 2/3 of the corn mixture into the prepared skillet. Layer with the ground chuck mixture and 1 cup of the cheese. Top with the remaining corn mixture and sprinkle with the remaining 1 cup cheese. Bake at 400 degrees for 30 to 45 minutes or until a wooden pick inserted in the center comes out clean.

Serves 8

Cajun Eggplant Dressing

1 pound eggplant, peeled, sliced
2 cups water
1 teaspoon salt
1/2 cup chopped celery
1/2 cup chopped onion
1/2 cup chopped green bell pepper
2 tablespoons butter
3 cups cooked rice
4 ounces crawfish tails, peeled
1/4 cup dry bread crumbs
1/4 cup chopped fresh parsley
2 teaspoons salt
1/4 teaspoon garlic powder
1/4 teaspoon black pepper
1/8 teaspoon cayenne pepper
1/4 teaspoon paprika

Combine the eggplant, water and 1 teaspoon salt in a saucepan. Bring to a boil; reduce the heat. Simmer for 5 minutes; drain.

Sauté the celery, onion and bell pepper in the butter in a skillet until the vegetables are tender-crisp. Stir in the eggplant, rice, crawfish tails, bread crumbs, parsley, 2 teaspoons salt, garlic powder, black pepper and cayenne pepper.

Spoon the eggplant mixture into a buttered 2-quart baking dish. Sprinkle with the paprika. Bake at 375 degrees for 25 to 30 minutes.

Serves 6

Rice Casserole

1/2 cup chopped green onions with tops
1/4 cup chopped bell pepper
1 cup sliced fresh mushrooms
1 1/4 cups rice
1 (8-ounce) can sliced water chestnuts,
drained
1 (10-ounce) can French onion soup

Sauté the green onions and bell pepper in a skillet sprayed with nonstick cooking spray. Stir in the mushrooms. Cook until the mushrooms are tender. Stir in the rice and water chestnuts.

Cook for 2 to 3 minutes, stirring frequently. Spoon the rice mixture into a 3-quart baking dish. Combine the soup with enough water to measure 2 1/2 cups. If using brown rice increase the liquid to 3 cups. Stir the soup mixture into the rice mixture. Bake at 350 degrees for 1 hour; brown rice may require a slightly longer cooking time.

Serves 8

Green Tomato Pickles

2 gallons chopped green tomatoes
1/2 gallon chopped white onions
2 quarts white vinegar
6 cups sugar
1 quart chopped hot chile peppers
1/2 cup salt
2 tablespoons pepper

Combine the green tomatoes, onions, vinegar, sugar, hot chile peppers, salt and pepper in a stockpot and mix well. Bring to a boil. Boil for 5 minutes, stirring occasionally.

Ladle the hot mixture into 14 hot sterilized pint jars, leaving a 1/2 inch headspace. Seal with 2-piece lids. Process in a boiling water bath for 10 minutes.

Makes 14 pints

Rice production in Louisiana ranks third in the nation. Rice is not just rice anymore. It is the main ingredient in many first-rate recipes…from salads to side dishes to entrées to desserts. And rice is not only tasty but also very healthy, with essential complex carbohydrates, vitamins, minerals, and protein, but stopping short on calories and fat.

Serving Louisiana

Bread, Breakfast
& Brunch

Bread, Breakfast & Brunch

Sweet Potato Biscuits

2 cups flour
2 tablespoons baking powder
1 teaspoon salt
2/3 cup shortening
1 cup mashed cooked sweet potatoes
6 tablespoons milk

Combine the flour, baking powder and salt in a bowl and mix well. Cut in the shortening until crumbly. Stir in the sweet potatoes and milk.

Knead the dough lightly on a floured surface. Roll the dough 1/2 inch thick and cut with a biscuit cutter. Arrange the rounds on a lightly greased baking sheet. Bake at 425 degrees for 15 to 20 minutes or until light brown.

Makes 1 dozen biscuits

Cowboy Coffee Cake

2 1/2 cups flour
2 cups packed brown sugar
1/2 teaspoon salt
2/3 cup shortening
2 teaspoons baking powder
1/2 teaspoon baking soda
1/2 teaspoon cinnamon
1/2 teaspoon nutmeg
1 cup sour milk
2 eggs, beaten

Combine the flour, brown sugar and salt in a bowl and mix well. Cut in the shortening until crumbly. Reserve 1/2 cup of the flour mixture. Stir the baking powder, baking soda, cinnamon and nutmeg into the remaining flour mixture. Add the milk and eggs and mix well.

Spoon the batter into 2 greased and floured 8-inch baking pans. Sprinkle with the reserved flour mixture. Bake at 375 degrees for 25 to 30 minutes or until the coffee cakes test done.

Serves 12

Cranberry Coffee Cake

Coffee Cake

2 cups flour
1 teaspoon baking powder
1 teaspoon baking soda
1/2 teaspoon salt
1 cup (2 sticks) butter, softened
1 cup sugar

2 eggs
1 cup sour cream
Vanilla extract to taste
1 (16-ounce) can whole cranberry sauce
1/2 cup chopped walnuts or pecans

Glaze

1/3 cup confectioners' sugar
5 teaspoons warm water

1/2 teaspoon almond extract

For the coffee cake, mix the flour, baking powder, baking soda and salt together. Beat the butter and sugar in a mixing bowl until creamy. Add the eggs and beat until blended. Add the flour mixture alternately with the sour cream, mixing well after each addition. Beat in the vanilla.

Spoon half the batter into a greased 9×9-inch baking pan. Layer with the cranberry sauce and walnuts. Top with the remaining batter. Bake at 350 degrees for 55 to 60 minutes or until a wooden pick inserted near the center comes out clean. Cool in pan on a wire rack.

For the glaze, combine the confectioners' sugar, warm water and flavoring in a bowl and mix well. Drizzle over the coffee cake.

Serves 9

Gingerbread

1 cup cane syrup	1 egg
1 teaspoon baking soda	1/2 teaspoon ginger
1 cup flour	1/2 teaspoon ground cloves
1/2 cup milk	

Combine the cane syrup and baking soda in a mixing bowl and mix well. Add the flour, milk, egg, ginger and cloves. Beat at medium speed for 3 minutes.

Spoon the batter into a greased and floured 8×8-inch baking pan. Bake at 325 degrees for 45 minutes or until the gingerbread springs back when lightly touched. Serve warm for breakfast or as a dessert with whipped cream and applesauce. Add additional spices as desired to the batter.

Serves 9

Beer Bread

3 cups self-rising flour	1 teaspoon salt (optional)
3 tablespoons sugar	1 (12-ounce) can beer
1 tablespoon baking powder	

Combine the self-rising flour, sugar, baking powder and salt in a bowl and mix well. Stir in the beer. Spoon the batter into a 5×9-inch loaf pan sprayed with nonstick cooking spray or brush the bottom and sides of the pan with 1 teaspoon of canola oil.

Bake at 375 degrees for 1 hour. Cool in pan for 5 to 10 minutes. Remove to a wire rack to cool completely.

Serves 12

Old-Fashioned Louisiana Corn Bread

1 cup white cornmeal
1/2 cup flour
1 1/2 teaspoons salt
1 teaspoon sugar (optional)
3 tablespoons corn oil

1/2 cup boiling water
1/2 cup milk
1 egg
1 1/2 teaspoons baking powder

Preheat a greased 9×9-inch baking pan in a 425-degree oven. Combine the cornmeal, flour, salt and sugar in a heatproof bowl and mix well. Stir in the corn oil. Add the boiling water and mix well. Stir in the milk. Cool to room temperature. Add the egg and baking powder to the batter and beat briskly. Spoon the batter into the prepared pan. Bake at 425 degrees for 20 minutes or until brown.

Serves 12

Raisin Bran Muffins

3/4 cup skim milk
1 cup 100% bran cereal
1 cup flour
1/4 cup sugar
2 1/2 teaspoons baking powder
1/4 teaspoon salt

1/2 cup chopped walnuts
1/2 cup chocolate-covered raisins
1/4 cup vegetable oil
1 egg, lightly beaten

Pour the skim milk over the cereal in a bowl. Let stand until soft. Sift the flour, sugar, baking powder and salt into a bowl and mix well. Stir in the walnuts and raisins. Stir the oil and egg into the cereal mixture. Add the cereal mixture to the flour mixture, stirring just until moistened. Fill 12 greased muffin cups 2/3 full. Bake at 400 degrees for 30 minutes.

Makes 1 dozen muffins

Lemon Bread

1¹/2 cups sifted flour
1 teaspoon baking powder
¹/8 teaspoon salt
1 cup sugar
¹/2 cup (1 stick) butter, softened
2 eggs

Grated zest of 1 lemon
¹/2 cup milk
1 cup chopped pecans
¹/2 cup confectioners' sugar, sifted
Juice of 1 lemon

Sift the flour, baking powder and salt into a bowl and mix well. Beat the sugar and butter in a mixing bowl until light and fluffy. Add the eggs and lemon zest and beat until blended. Add the dry ingredients and milk alternately ¹/2 at a time, beating well after each addition. Fold in the pecans.

Spoon the batter into a greased and floured 4×8-inch loaf pan. Bake at 350 degrees for 45 minutes. Pierce the top of the hot loaf in several places with a wooden pick. Combine the confectioners' sugar and lemon juice in a bowl and mix well. Drizzle the lemon glaze over the hot loaf. Let stand for 10 minutes. Remove to a wire rack to cool completely.

Serves 12

Why buy a lemon? Think of the best cooks you know. You'll discover they are never without fresh lemons. Fresh lemons bring out the best in other foods. The zesty flavor of the peel and juice adds a tang and tartness to otherwise humdrum foods. Select lemons with smooth firm skins, free of blemishes and soft spots. A lemon yields more juice when left at room temperature and rolled on the counter before squeezing.

Lemon Blueberry Bread

Bread

1¹/2 cups flour
1 teaspoon baking powder
¹/8 teaspoon salt
6 tablespoons butter

1 cup sugar
2 eggs
¹/2 cup milk

2 teaspoons grated
lemon zest
1 cup fresh blueberries
1 tablespoon flour

Lemon Glaze

3 tablespoons lemon juice

¹/3 cup sugar

For the bread, mix 1¹/2 cups flour, baking powder and salt together. Beat the butter in a mixing bowl until creamy. Add the sugar gradually, beating constantly at medium speed until blended. Add the eggs 1 at a time, beating well after each addition. Add the dry ingredients alternately with the milk, beginning and ending with the dry ingredients and mixing well after each addition. Stir in the lemon zest.

Toss the blueberries with 1 tablespoon flour in a bowl. Fold the blueberries into the batter. Spoon the batter into a greased 4×8-inch loaf pan. Bake at 350 degrees for 55 minutes or until a wooden pick inserted in the center comes out clean.

For the glaze, combine the lemon juice and sugar in a saucepan and mix well. Cook until the sugar dissolves, stirring frequently. Pierce the top of the hot loaf in several places with a wooden pick. Pour the lemon glaze over the hot loaf. Cool in pan for 30 minutes. Remove to a wire rack to cool completely.

Serves 12

Photograph for this recipe appears on page 151.

Gifts of food from your kitchen are truly gifts of love. When you make it yourself, no matter how simple, you are giving of yourself. And in this day of high prices, it is the only sensible way to show you care without breaking your already strained budget. Whether making miniature loaves of bread, candy, cookies, or tea mixes, be creative in putting together a gift basket for a special friend.

Anadama Bread

6¹/4 to 6³/4 cups flour
¹/2 cup cornmeal
2 envelopes dry yeast
2 cups water

¹/2 cup dark molasses
¹/3 cup shortening
1 tablespoon salt
2 eggs, beaten

Mix 3 cups of the flour, cornmeal and yeast in a mixing bowl. Combine the water, molasses, shortening and salt in a saucepan. Heat just to 115 to 120 degrees or until lukewarm, stirring constantly. Stir the molasses mixture into the flour mixture. Add the eggs. Beat at low speed for 30 seconds. Beat at high speed for 3 minutes, scraping occasionally. Stir in as much of the remaining flour as possible with a wooden spoon.

Place the dough on a lightly floured surface. Knead for 3 to 5 minutes, adding enough of the remaining flour to make a medium-soft dough. Shape the dough into a ball and place in a lightly greased bowl, turning to coat the surface. Let rise, covered, in a warm place for 1 to 1¹/4 hours or until doubled in bulk. Punch the dough down. Divide into 2 equal portions. Let rest, covered, for 10 minutes.

Shape each portion into a loaf in a greased 5×9-inch loaf pan. Let rise, covered, for 45 to 60 minutes or until almost doubled in bulk. Bake at 375 degrees for 35 to 40 minutes or until the loaves test done. Cool in pans for 10 minutes. Remove to a wire rack to cool completely.

Serves 24

No-Knead Maple Molasses Bran Bread

1¹/4 cups skim milk
¹/4 cup (¹/2 stick) margarine
1¹/4 cups 100% bran cereal
1 envelope dry yeast
¹/4 cup warm water
(105 to 115 degrees)
1 teaspoon plus 2 tablespoons
maple syrup

3 tablespoons molasses
1 egg, beaten
2 cups all-purpose flour
3/4 cup whole wheat flour
1 teaspoon lite salt

Combine the skim milk and margarine in a saucepan. Bring to a boil over medium heat, stirring frequently. Pour the skim milk mixture over the cereal in a mixing bowl and stir. Let stand for 30 minutes.

Dissolve the yeast in the warm water in a small bowl. Stir in 1 teaspoon of the maple syrup. Let stand for 15 minutes or until bubbly. Stir the maple syrup mixture into the cereal mixture. Add the remaining 2 tablespoons maple syrup, molasses, egg, 3/4 cup of the all-purpose flour, whole wheat flour and lite salt.

Beat at medium speed for 3 minutes, scraping the bowl occasionally. Stir in the remaining 1¹/4 cups all-purpose flour until blended. Spoon into a greased 2-quart round baking dish. Let rise, covered with plastic wrap, in a warm place for 45 to 60 minutes or until the dough rises to the top edge of the baking dish. Bake at 350 degrees for 45 minutes, covering loosely with foil after 30 minutes. Remove the loaf to a wire rack to cool. May be frozen for future use. You may substitute ¹/4 cup egg substitute for the egg.

Serves 32

Classic Cloverleaf
Yeast Rolls

2 envelopes dry yeast	1 cup sugar
3/4 cup lukewarm water	1 teaspoon salt
1 cup boiling water	2 eggs
1 cup shortening	6 cups sifted flour

Dissolve the yeast in the lukewarm water in a small bowl and mix well. Combine the boiling water and shortening in a heatproof mixing bowl and stir until the shortening melts. Add the sugar and salt and stir until dissolved. Beat in the eggs. Stir in the yeast mixture.

Add the flour 1 cup at a time, beating after each addition until the dough adheres and is easily handled. Shape the dough into a ball and place in a floured sealable plastic bag; seal tightly. Chill for 8 to 10 hours.

Spray muffin cups with nonstick cooking spray. Shape the dough into 1-inch balls. Arrange 3 balls in each prepared muffin cup. Let rise for 2$1/2$ to 3 hours. Bake at 375 degrees for 10 to 15 minutes or until light brown. Serve warm.

Makes 30 rolls

Breads are a wonderful way of upgrading the nutritional value of your diet. For example, increase calcium in your diet by incorporating cheese, milk, and yogurt in your favorite bread recipes. Calcium-rich breads can be an important source of this vital mineral and a flavorful addition to your diet.

Refrigerator Rolls

2 envelopes dry yeast
$1/2$ cup lukewarm water
$1^1/2$ cups milk
$1/2$ cup sugar
$1/4$ cup vegetable oil or
shortening

2 teaspoons salt
6 cups (about) flour
1 egg, beaten
Melted margarine (optional)

Dissolve the yeast in the lukewarm water in a bowl and mix well. Heat the milk in a saucepan. Stir in the sugar, oil and salt. Let stand until lukewarm. Add 2 cups of the flour to the milk mixture and mix well. Stir in the yeast mixture and egg. Add just enough of the remaining flour to make a soft dough and mix well.

Knead the dough on a lightly floured surface for 5 to 10 minutes or until smooth and elastic. Shape the dough into a ball and lightly grease the outer surface. Place the dough in a sealable plastic bag and seal tightly. Let rise until doubled in bulk. Punch the dough down and press the air out of the bag; seal tightly. Place the dough in the refrigerator for 8 to 10 hours. Remove the dough and punch down. Let rest for 10 to 20 minutes. Shape as desired and place the rolls in a greased baking pan.

Let rise for 45 minutes or until doubled in bulk. Bake at 425 degrees for 8 to 12 minutes or until light brown. Brush with melted margarine. Use also as a basic recipe for coffee cakes; makes approximately three.

Makes 5 dozen rolls.

Bread stored at room temperature stays fresh longer. Bread stored in the refrigerator becomes stale faster but will not mold as quickly. You may freeze bread for up to six months. Store rice, flour, noodles, and cornmeal in tightly sealed containers in a dry environment. Avoid rinsing rice, cooked spaghetti, and noodles when possible as the rinsing process removes important vitamins.

Whole Wheat Refrigerator Rolls

2 envelopes dry yeast
1/2 cup lukewarm water
4 cups milk
1 cup (2 sticks) butter or
margarine
1 cup sugar

3 cups whole wheat flour
2 tablespoons salt
All-purpose flour
2 teaspoons (slightly heaping)
baking powder
1 teaspoon baking soda

Dissolve the yeast in the lukewarm water in a bowl. Scald 1 cup of the milk, butter and sugar in a saucepan. Stir the milk mixture into the remaining 3 cups milk in a mixing bowl. Let stand until lukewarm. Stir in the yeast mixture.

Combine 2 cups of the whole wheat flour and salt in a bowl and mix well. Stir the whole wheat flour mixture into the yeast mixture. Beat until blended, adding enough all-purpose flour until the consistency of a thick batter. Let rise in a warm place for about 1 hour, checking frequently. Punch the dough down.

Sift the remaining 1 cup whole wheat flour, baking powder and baking soda into a bowl and mix well. Add the whole wheat flour mixture to the dough. Knead for 10 minutes, adding all-purpose flour as needed. Place the dough in a greased bowl, turning to coat the surface. Chill, covered, in the refrigerator for up to 2 weeks.

Pinch off the desired amount of dough as needed and shape into rolls. Arrange the rolls in a greased baking pan. Let rise in a warm place for 1 to 1 1/2 hours. Bake at 425 degrees for 10 minutes. Serve warm. You may substitute butter-flavor shortening for the butter or margarine.

Makes 80 to 100 small rolls

Flensjes

1 cup flour
1 egg
1 1/4 cups milk, water or
reconstituted nonfat
dry milk

1 tablespoon sugar
Vegetable oil
1 to 2 tablespoons brown
sugar, syrup or molasses

Combine the flour and egg in a bowl and mix well. Add the milk gradually, stirring constantly until smooth. Stir in the sugar.

Add enough oil to a cast-iron skillet to cover the bottom. Heat over medium heat. Test heat of skillet by flicking a drop of water into the skillet. The water should dance if the oil is the correct temperature. Pour a circle of the batter into the skillet, tilting the skillet to allow the batter to cover the bottom.

Cook until bubbles appear on the top and turn. The cooked side should appear white with brown flecks. Cook the second side until white with brown flecks. To serve, place the flensje on a plate. Spoon brown sugar down the middle or drizzle syrup or molasses down the middle. Roll up the flensje from bottom to top and turn 90 degrees. Cut into bite-size slices. Use multiple skillets if cooking for more than one.

Serves 1

Flensjes (flens-yuhs) is a common Dutch recipe. Although very similar to crepes, flensjes are heavier because they are cooked with more oil. Flensjes are intended as a high energy food for working people in cold climates. This makes them particularly attractive as a breakfast on camping trips. In the logging camps of Canada, an identical recipe is cooked up into six-inch-wide pancakes and called "Finlander Pancakes."

Overnight French Toast

1 (13- to 14-inch long) loaf soft-crust
French or Italian bread
1/4 cup (1/2 stick) butter or margarine, melted
1 2/3 cups milk
3 tablespoons sugar
2 eggs, lightly beaten
1/2 teaspoon vanilla extract
1/8 teaspoon salt
1/8 teaspoon cinnamon
1/8 teaspoon nutmeg
2 tablespoons sugar

Trim the ends of the loaf and discard. Cut the loaf into twelve 1-inch slices. Brush 1 side of each bread slice with melted butter. Arrange the slices butter side up in a 9×13-inch baking dish sprayed with nonstick cooking spray. The slices may be wedged in to fit snugly in the baking dish.

Whisk the milk, 3 tablespoons sugar, eggs, vanilla, salt, cinnamon and nutmeg in a bowl. Pour the milk mixture over the bread. Chill, covered with plastic wrap, for 1 to 10 hours or until the bread has absorbed the milk mixture. Sprinkle with 2 tablespoons sugar. Bake at 425 degrees for 20 to 25 minutes or until golden brown.

Serves 6

Make-Ahead Breakfast Bake

12 eggs
1/2 cup milk
1/2 teaspoon salt
1/4 teaspoon pepper
1 tablespoon butter
1 cup sour cream
12 slices bacon, crisp-cooked, crumbled
1 cup shredded sharp Cheddar cheese

Whisk the eggs in a bowl until blended. Add the milk, salt and pepper and mix well. Heat the butter in a large skillet over medium-low heat. Add the egg mixture to the hot skillet.

Cook until the eggs are set but still moist, stirring occasionally. Remove from heat. Let stand until cool. Stir in the sour cream. Spread the egg mixture in a buttered shallow 2-quart 7×12-inch baking dish. Sprinkle with the bacon and cheese.

Chill, covered with foil, for 8 to 10 hours; remove the cover. Bake at 300 degrees for 15 to 20 minutes or until bubbly and heated through. Store leftovers in the refrigerator.

Serves 8

Eggs Fantastic

1 pound sausage, crumbled
4 ounces fresh mushrooms, chopped
1 onion, chopped
1/4 teaspoon oregano
Salt and pepper to taste
6 eggs
3 tablespoons sour cream
1/8 teaspoon Tabasco sauce

6 tablespoons Mexican tomato and green chile sauce
8 ounces Cheddar cheese, shredded
8 ounces Velveeta cheese, shredded
8 ounces mozzarella cheese, shredded

Brown the sausage with the mushrooms and onion in a skillet, stirring until the sausage is crumbly; drain. Stir in the oregano, salt and pepper. Spoon the sausage mixture into a greased 9×13-inch baking dish.

Combine the eggs, sour cream and Tabasco sauce in a blender. Process for 15 seconds. Pour the egg mixture over the prepared layer. Bake at 400 degrees for 10 minutes or until set. Spread the tomato and green chile sauce over the top. Sprinkle with a mixture of the Cheddar cheese, Velveeta cheese and mozzarella cheese. Broil until bubbly. Serve immediately.

Serves 6

Store eggs in the carton in the refrigerator for optimal flavor and freshness. If stored uncovered (for instance, the egg bin in your refrigerator) they lose moisture and can absorb odors from other foods. Use eggs within five weeks of purchase.

An LSU AgCenter scientist has discovered a way to get chickens to lay more eggs. After a dose of Eggmax, which is the registered trademark of the vaccine, hens can lay as many as three dozen more eggs during their reproductive lives. Considering there are tens of millions of chickens in the United States, that's a lot of extra eggs!

Brunch Casserole

1 (8-count) can crescent rolls
1 pound bulk sausage, crumbled
2 cups shredded mozzarella cheese
3/4 cup milk
4 eggs, beaten
1/4 teaspoon salt
1/8 teaspoon pepper

Unroll the crescent roll dough. Pat the dough over the bottom of a buttered 9×13-inch baking dish, pressing the perforations and edges to seal. Brown the sausage in a skillet over medium heat, stirring until crumbly; drain. Sprinkle the sausage and cheese over the dough.

Whisk the milk, eggs, salt and pepper in a bowl until blended. Pour over the prepared layers. Bake at 425 degrees for 15 minutes or until set. Let stand for 5 minutes. Cut into squares and serve immediately.

Serves 8

Cajun Eggs

1 pound smoked sausage
1/4 cup chopped onion
1/4 cup chopped green onions
8 eggs, beaten
1 cup shredded Cheddar cheese
1/4 teaspoon salt
1/4 teaspoon Cajun seasoned salt
Red pepper to taste
Black pepper to taste
Tabasco sauce to taste
1 tablespoon flour
1/2 cup to 3/4 cup milk

Cut the sausage into 1/2-inch slices. Pan-fry the sausage in a skillet until cooked through. Drain, reserving the pan drippings. Sauté the onion and green onions in a nonstick skillet until tender. Combine the sautéed onion mixture, eggs, cheese, salt, Cajun seasoned salt, red pepper, black pepper and Tabasco sauce in a bowl and mix well.

Make a roux using the reserved pan drippings and flour. Cook until thickened and light brown in color, stirring constantly. Add enough of the milk to the roux to make of the consistency of a thin gravy, stirring constantly. Stir in the sausage and egg mixture. Cook until the mixture is of the consistency of scrambled eggs, stirring constantly.

Serves 8

Corny Egg Casserole

2 (17-ounce) cans
cream-style corn
1 (6-ounce) package Mexican
corn bread mix

1/2 cup vegetable oil
6 eggs, beaten
2 cups shredded Cheddar
cheese

Combine the corn, corn bread mix, oil and eggs in a bowl and mix well. Spoon the corn mixture into a 9×13-inch baking dish. Bake at 350 degrees for 30 to 45 minutes or until set and brown. Sprinkle with the cheese. Bake just until the cheese melts.

Serves 10

Breakfast Pizza

1 pound bulk pork sausage
1 (8-count) can crescent rolls
1 cup loosely packed frozen
hash brown potatoes
1 cup shredded Cheddar
cheese

4 eggs, beaten
1/4 cup milk
1/2 teaspoon salt
1/4 teaspoon pepper
2 tablespoons grated Parmesan
cheese (optional)

Brown the sausage in a skillet, stirring until crumbly; drain. Unroll the crescent roll dough and separate into 8 triangles. Arrange the triangles in a circle with points facing the center on a greased 12- to 15-inch pizza pan. Press the dough over the bottom and up the side of the pan to form a crust. Spoon the sausage over the crust. Sprinkle with the hash brown potatoes and then with the Cheddar cheese.

Whisk the eggs, milk, salt and pepper in a bowl until blended. Pour over the top. Bake at 375 degrees for 25 minutes. Sprinkle with the Parmesan cheese. Bake for 5 minutes longer.

Serves 8

Serving Louisiana
Sweet Treats

Sweet Treats

Bananas Foster

1/2 cup (1 stick) butter
1 cup packed brown sugar
3 tablespoons crème de banane
1/2 teaspoon cinnamon
4 medium bananas, sliced lengthwise
1/4 cup rum
3 cups vanilla ice cream

Heat the butter in a skillet over medium heat. Stir in the brown sugar, crème de banane and cinnamon. Cook until the brown sugar dissolves and the mixture is bubbly, stirring frequently. Stir in the sliced bananas.

Cook just until the bananas are tender and beginning to glaze, stirring occasionally. Remove from heat. Pour the rum into a microwave-safe cup. Microwave until warm. Pour over the bananas. Spoon the banana mixture over the ice cream in dessert bowls.

Serves 6

Chocolate Turtle Cheesecake

1 (14-ounce) package caramels
1 (5-ounce) can evaporated milk
1 cup chopped pecans
1 (9-inch) chocolate graham cracker pie shell
3 ounces cream cheese, softened
1/2 cup sour cream
1/4 cup whole milk
1 (4-ounce) package chocolate instant pudding mix
1/2 cup fudge ice cream topping
1/4 cup chopped pecans

Combine the caramels and evaporated milk in a heavy saucepan. Cook over medium-low heat for 5 minutes or until blended, stirring constantly. Stir in 1 cup pecans. Pour the caramel mixture into the pie shell.

Combine the cream cheese, sour cream and whole milk in a blender. Process until smooth. Add the pudding mix. Process for 30 seconds or until blended. Spread the cream cheese mixture over the prepared layer. Chill, loosely covered, for 15 minutes or until set.

Drizzle the fudge topping over the top in a decorative pattern. Sprinkle with 1/4 cup pecans. Chill, loosely covered, until serving time.

Serves 8

Favorite Peach Cobbler

Pastry

2 cups sifted flour
1 teaspoon salt

2/3 cup shortening
1/2 cup milk

Peach Filling

6 large peaches, peeled, sliced
1 cup sugar
1 cup water
1/4 teaspoon freshly
grated nutmeg

1/4 teaspoon salt
1/4 cup (1/2 stick) margarine
1/4 teaspoon vanilla extract
1/4 cup (1/2 stick) margarine,
melted

For the pastry, combine the flour and salt in a bowl and mix well. Cut in the shortening until crumbly. Add the milk and stir until the mixture forms a ball. Roll the pastry on a lightly floured surface. Cut into strips.

For the filling, combine the peaches, sugar, water, nutmeg and salt in a saucepan. Cut a few of the pastry strips and add to the peach mixture. Simmer for a few minutes, stirring occasionally. Stir in 1/4 cup margarine and vanilla. Pour the peach mixture into a baking dish. Arrange the pastry strips lattice-fashion over the top. Drizzle with 1/4 cup melted margarine.

Bake at 375 degrees for 30 minutes or until golden brown.

Serves 8

Ripeness is a measure of quality when selecting peaches. A red color, or "blush," is not always a sign of ripeness. The skin color between the red areas indicates the peach's ripeness. It should be yellow or creamy, not green. Avoid peaches that are very firm or hard with a green ground color because they probably will not ripen properly. Peaches are a good source of niacin and potassium, and contain small amounts of iron and other minerals and vitamins.

Bread Pudding with Bourbon Sauce

Bread Pudding

1 large loaf French bread
3 ounces raisins
3 quarts 2% milk
1 cup bourbon
12 eggs

1½ cups sugar
4 teaspoons vanilla extract
4 or 5 drops of yellow food
  coloring

Bourbon Sauce

⅓ to ½ cup sugar
2½ tablespoons cornstarch
1 cup water
¼ cup (½ stick) butter

3 or 4 tablespoons bourbon
½ teaspoon vanilla extract
4 drops of yellow food coloring

For the pudding, spray a 10×15-inch baking dish lightly with butter-flavor nonstick cooking spray. Tear the bread into 1-inch pieces and arrange in the prepared baking dish. Sprinkle with the raisins.

Process the 2% milk, bourbon, eggs, sugar, vanilla and food coloring in batches in a blender until smooth. Pour the milk mixture over the prepared layers. Let stand for 5 to 10 minutes.

Bake at 350 degrees for 1¼ hours or until a knife inserted in the center comes out clean. Cut into 2½-inch squares. Pour hot Bourbon Sauce over the squares in the baking dish. Serve warm. Add chopped fresh apples, chopped dried fruit or 1 cup freshly grated coconut for variety.

For the sauce, combine the sugar and cornstarch in a saucepan and mix well. Stir in the water. Cook until thickened, stirring constantly. Remove from heat. Stir in the butter, bourbon, vanilla and food coloring.

To serve, cut the warm bread pudding into 2½-inch squares. Drizzle some of the hot sauce over each serving.

Serves 24

Death by Chocolate

1 (2-layer) package chocolate cake mix
3/4 cup Kahlúa
3 (4-ounce) packages chocolate instant
pudding mix
5 cups milk
24 ounces whipped topping
6 (1-ounce) Heath candy bars, crushed

Prepare and bake the chocolate cake using package directions for a 9×13-inch cake pan. Pierce the top of the hot cake with a fork and drizzle with the Kahlúa. Cool in pan on a wire rack. Crumble the cake. Prepare the pudding mixes with the milk using package directions.

Layer the cake, pudding, whipped topping and candy 1/2 at a time in a large trifle or glass bowl. Chill, covered, until serving time.

Serves 25

Guiltless Peach Dessert

1 large package sugar-free vanilla pudding mix
3 cups skim milk
8 ounces nonfat cream cheese, softened
1 cup peach yogurt
2 to 3 teaspoons sugar-free peach
gelatin granules
1 angel food cake, torn into bite-size pieces
6 peaches, peeled, sliced
8 ounces reduced-fat whipped topping

Prepare the pudding mix with the skim milk using package directions. Beat in the cream cheese, yogurt and peach gelatin.

Layer the angel food cake, pudding mixture and peaches 1/2 at a time in a glass serving bowl. Spread with the whipped topping. Chill, covered, until serving time. The flavor is enhanced if prepared 1 day in advance and stored, covered, in the refrigerator.

Serves 12

Pumpkin Roll

Cake

1 cup sugar
3/4 cup flour
1 teaspoon baking soda
1 teaspoon cinnamon

2/3 cup canned pumpkin
3 eggs, lightly beaten
Confectioners' sugar to taste

Cream Cheese Filling

8 ounces cream cheese,
  softened
1/4 cup (1/2 stick) butter,
  melted

1 cup confectioners' sugar
1/2 teaspoon vanilla extract
1/2 cup chopped nuts
  (optional)

For the cake, combine the sugar, flour, baking soda and cinnamon in a bowl and mix well. Stir in the pumpkin and eggs. Spread the pumpkin mixture on a greased and floured 9×12-inch baking sheet with sides. Bake at 375 degrees for 15 minutes. Cool on baking sheet for 10 to 15 minutes.

Dust a clean tea towel generously with confectioners' sugar. Invert the cake onto the tea towel. Roll the warm cake in the tea towel as for a jelly roll from the short side and place on a wire rack to cool. Unroll the cooled cake carefully and remove the tea towel.

For the filling, beat the cream cheese and butter in a mixing bowl until creamy. Add the confectioners' sugar and vanilla. Beat until the filling is a spreading consistency, scraping the bowl occasionally. Stir in the nuts.

Spread the cream cheese mixture to the edge and reroll. Wrap the roll in foil. Chill in the refrigerator until serving time.

Serves 15

*The secret to making a jelly roll cake is rolling it while
the cake is warm.*

Grandmother's Favorite Butter Tarts

1 (17-ounce) package
puff pastry
2 cups packed brown sugar
1/4 cup (1/2 stick) butter,
melted
1 cup raisins, minced

1/4 cup figs, minced
1/4 cup dried nectarines,
minced
1 cup pecan pieces, toasted
2 eggs
2 teaspoons vanilla extract

Thaw the puff pastry using package directions. Combine the brown sugar and butter in a bowl and mix well. Stir in the raisins, figs, nectarines and pecans. Whisk the eggs and vanilla in a bowl until blended. Add the egg mixture to the fruit mixture and mix well.

Place the puff pastry on a lightly floured surface. Roll with a rolling pin to smooth any creases. Cut twenty-four 2 1/2-inch rounds with a biscuit cutter. Pat each round over the bottom and up the side of a buttered and floured miniature muffin cup.

Fill the muffin cups 3/4 full with the fruit mixture. Bake at 350 degrees for 15 to 20 minutes or until golden brown. Serve warm or at room temperature. You may prepare up to 2 days in advance and store in a container with a tight-fitting lid. Substitute dried apricots or dried figs for the raisins if desired.

Makes 2 dozen tarts

Figs are one of the most perishable fresh fruits. Pick when fully ripe, handle carefully, and use quickly because they spoil so readily. Wear gloves when picking or peeling figs to protect your hands from the irritating juices. Figs are a good source of fruit sugars for energy. They are also a good source of iron, the B vitamins and niacin, as well as containing fair amounts of calcium and other minerals. Figs are a good source of dietary fiber.

Raisin Tarts

Pastry

1 cup shortening
1/2 teaspoon salt

1/2 cup boiling water
1 1/2 cups flour

Raisin Filling

2 cups sugar
1 cup (2 sticks) butter or
margarine, softened
2 teaspoons vanilla extract

4 egg yolks, beaten
4 egg whites, beaten
1 1/2 cups raisins

For the pastry, combine the shortening and salt in a heatproof bowl and mix well. Stir in the boiling water. Let stand until cool. Add the flour and mix well. Shape the pastry into a ball and place in a bowl. Chill, covered, for 8 to 10 hours.

Roll the pastry thin on a lightly floured surface. Cut 24 rounds with a biscuit cutter. Pat each round over the bottom and up the side of a muffin cup.

For the filling, beat the sugar and butter in a mixing bowl until creamy. Stir in the vanilla. Beat in the egg yolks until blended. Add the egg whites and beat until smooth. Stir in the raisins. Spoon the raisin mixture into the prepared muffin cups. Bake at 350 degrees for 30 to 35 minutes or until light brown.

Makes 2 dozen tarts

Fruit Sherbet

6 bananas, mashed
2 cups sugar
1 (20-ounce) can crushed
pineapple, drained
1 cup shredded coconut
(optional)

1 (17-ounce) can apricots,
drained, chopped
2 (6-ounce) cans frozen
orange juice concentrate,
partially thawed

2 juice cans water
Juice of 3 lemons
1 (1-liter) bottle ginger ale,
chilled

Combine the bananas, sugar, pineapple, coconut, apricots, orange juice concentrate, water and lemon juice in a bowl and mix well. Pour the fruit mixture into ice trays or a freezer container. Freeze until firm.
 To serve, scoop the sherbet into parfait glasses. Pour enough ginger ale over each serving to fill the glass.

Serves 20

Cantaloupe Sorbet

1 cantaloupe or other
muskmelon, cubed

3/4 cup sugar
1 teaspoon lemon juice

Combine the cantaloupe, sugar and lemon juice in a food processor. Process until smooth. Pour the cantaloupe mixture into an ice cream freezer container. Freeze using manufacturer's directions.

Serves 8

Picking a good cantaloupe is always a challenge. There should be no trace of a stem and there should be a very definite pronounced cavity where the melon was pulled from the vine. Also, look for a thick netting with a slight golden background color. Cantaloupes should be shaped like a football and have a delicate melon aroma. They are rich in vitamins A and C.

Melon Ice

1 tablespoon unflavored
gelatin
3 cups melon purée
3/4 cup sugar
2 tablespoons lemon juice

1/2 teaspoon grated gingerroot,
or 1/8 teaspoon ground
ginger
10 ounces lemon-lime soda

Combine the gelatin with 1/4 cup of the melon purée in a saucepan and mix well. Let stand for 5 minutes or until softened. Cook over low heat until the gelatin dissolves, stirring frequently.

Combine the gelatin mixture, remaining melon purée, sugar, lemon juice, gingerroot and soda in a bowl and mix well. Pour the melon mixture into an ice cream freezer container. Freeze using manufacturer's directions. For the melon purée, use cantaloupe, honeydew melon or any other variety of muskmelon.

Serves 8

Selecting a good, sweet watermelon is always tricky. If you are a "thumper," listen for a dull, hollow sound. A better way is to turn the watermelon over…the underside should be yellow with a healthy sheen to the rind. And the watermelon should be heavy for its size. Watermelons are fat-free, cholesterol-free, low in sodium, high in vitamin C, and are a good source of vitamin A.

Vanilla Ice Cream

6 eggs
2¹/2 cups sugar
1 quart whole milk
1 (12-ounce) can
evaporated milk

2 tablespoons vanilla extract
2 pints half-and-half
Whole milk

Beat the eggs in a mixing bowl until foamy. Add the sugar gradually, beating constantly until mixed. Scald 1 quart whole milk and the evaporated milk in a saucepan. Stir a small amount of the hot milk into the egg mixture. Stir the egg mixture into the hot milk. Cook until the mixture coats a spoon, stirring frequently. Stir in the vanilla. Cool in the refrigerator. Stir in the half-and-half.

Pour the milk mixture into an ice cream freezer container. Add enough additional whole milk to fill the container 2/3 full. Freeze using manufacturer's directions. For variety, add 1 quart mashed fresh ripe fruit. If using a sweet fruit such as figs, reduce the sugar to 2 cups if desired.

Reduce the fat grams by substituting three 12-ounce cans evaporated skim milk for the evaporated milk and half-and-half and 1% milk for the whole milk.

Serves 16

The LSU AgCenter's Dairy Store and its accompanying creamery offer teaching and research opportunities for the Department of Dairy Science, as well as ice cream treats. The creamery, a small-scale production facility, is used to teach students to make ice cream and cheese and produces about 6,000 gallons of ice cream and 5,000 pounds of cheese per year. Operated almost exclusively by student labor, the store in its current location has been open since 1972. The first dairy store opened on campus in 1929.

Apple Fig Cake

2¹/2 cups flour
1 teaspoon cinnamon
1 teaspoon ground cloves
1 teaspoon allspice
¹/8 teaspoon salt
1¹/2 cups sugar
¹/2 cup (1 stick) margarine,
   softened

2 eggs
1¹/2 cups chopped apples,
   stewed, drained
2 teaspoons baking soda
1 cup golden raisins
1 cup pecan pieces
1 cup fig preserves, chopped

Line the bottom of a 9- to 10-inch springform pan with waxed paper. Mix the flour, cinnamon, cloves, allspice and salt together. Beat the sugar and margarine in a mixing bowl until creamy. Add the eggs 1 at a time, beating well after each addition.

Mix the apples and baking soda in a bowl. Add the apple mixture to the creamed mixture and stir just until mixed. Stir in the flour mixture. Fold in the raisins and pecans. Fold in the preserves.

Spoon the batter into the prepared pan. Bake at 300 degrees for 1¹/2 hours or until a wooden pick inserted in the center comes out clean. Remove the cake from the pan and discard the waxed paper. Cool on a wire rack.

Serves 20

Applesauce Cake

5 medium cooking apples,
peeled, chopped
4 cups flour
2 teaspoons baking soda
2 teaspoons cinnamon
2 teaspoons allspice
2 cups chopped pecans

2 cups raisins
1 cup (2 sticks) butter,
softened
2 cups sugar
2 eggs, beaten
1/2 teaspoon salt
2 teaspoons vanilla extract

Combine the apples with enough water to cover in a saucepan. Bring to a boil. Boil until tender; drain. Mash the apples in a bowl. Mix the flour, baking soda, cinnamon and allspice together. Toss the pecans and raisins with some of the flour mixture in a bowl until coated. Stir the pecan and raisin mixture into the flour mixture.

Beat the butter in a mixing bowl until creamy. Add the sugar gradually, beating constantly until blended. Beat in the eggs and salt. Add the flour mixture and vanilla and mix well. Spoon the batter into an angel food cake pan. Bake at 375 degrees for 1 hour or until a wooden pick inserted near the center comes out clean. When storing the cake, place an apple wedge in the middle to keep moist.

Serves 16

Blueberry Cake

8 ounces cream cheese,
softened
1/2 cup vegetable oil
3 eggs, or 6 egg whites

1 (2-layer) package butter-
flavor cake mix
2 cups fresh blueberries

Beat the cream cheese, oil and eggs in a mixing bowl until creamy, scraping the bowl occasionally. Stir in the cake mix. Fold in the blueberries.

Spoon the batter into a greased bundt pan. Bake at 350 degrees for 50 to 60 minutes or until the cake tests done. Cool in pan for 10 minutes. Remove to a wire rack to cool completely.

Serves 16

Blueberries increase in size and improve in flavor for several days after they turn blue. It will take three to six days after turning blue for them to become fully ripe. When choosing blueberries, be sure they are plump and firm. Keep refrigerated until ready for use. Freeze blueberries without rinsing to prevent the skins from becoming tough. The blueberry has bacteria-fighting properties that help prevent urinary tract infections.

Carrot Cake with Cream Cheese Frosting

Cake

2 cups flour
2 cups sugar
2 teaspoons baking powder
2 teaspoons cinnamon
1 1/2 teaspoons baking soda
1 teaspoon salt

1 1/2 cups vegetable oil
4 eggs
3 cups grated carrots
1 cup chopped pecans
1/2 to 1 cup raisins (optional)

Cream Cheese Frosting

1/2 cup (1 stick) butter or
margarine, softened
8 ounces cream cheese,
softened

1 (1-pound) package
confectioners' sugar
1 tablespoon vanilla extract

For the cake, combine the flour, sugar, baking powder, cinnamon, baking soda and salt in a bowl and mix well. Add the oil and eggs and mix until blended. Fold in the carrots, pecans and raisins.

Spoon the batter into 3 greased 9-inch cake pans or a bundt pan. Bake at 400 degrees until the layers test done. Cool in pans for 10 minutes. Remove to wire racks to cool completely.

For the frosting, beat the butter and cream cheese in a mixing bowl until creamy, scraping the bowl occasionally. Add the confectioners' sugar and vanilla. Beat until of a spreading consistency. Spread the frosting between the layers and over the top and side of the cake.

Serves 12

Carrot Cake with Orange Glaze

Cake

3 cups flour
2 teaspoons baking powder
1 teaspoon baking soda
1 teaspoon cinnamon
1/2 teaspoon salt
1 cup applesauce
1 cup sugar

1 cup packed light
   brown sugar
6 egg whites
Juice of 1 orange
3 cups shredded peeled carrots
1 cup raisins (optional)

Orange Glaze

2 cups confectioners' sugar
2 tablespoons applesauce
1 to 2 tablespoons orange juice

Chopped pecans or walnuts
   (optional)

For the cake, mix the flour, baking powder, baking soda, cinnamon and salt in a bowl. Combine the applesauce, sugar, brown sugar, egg whites and orange juice in a mixing bowl. Beat until blended. Add the flour mixture and beat until smooth. Stir in the carrots and raisins.

Spoon the batter into a greased and floured tube pan. Bake at 350 degrees for 60 to 65 minutes or until the cake tests done. Cool in pan for 10 minutes. Invert on a wire rack to cool completely.

For the glaze, combine the confectioners' sugar, applesauce and orange juice in a bowl and stir until of a drizzling consistency. Stir in the pecans. Drizzle the glaze over the cake.

Serves 16

Chocolate Cake

Cake

2 cups sugar
2 cups flour
1/2 cup baking cocoa
2 1/2 teaspoons baking soda
1/4 teaspoon salt

1 cup water
1 cup vegetable oil
1 cup buttermilk
2 eggs
1 teaspoon vanilla extract

Cream Cheese Frosting

4 ounces cream cheese,
softened
1/4 cup (1/2 stick) butter,
softened

1 (1-pound) package
confectioners' sugar
3 tablespoons milk
1/2 teaspoon vanilla extract

For the cake, mix the sugar, flour, baking cocoa, baking soda and salt in a mixing bowl. Add the water, oil, buttermilk, eggs and vanilla. Beat for 2 minutes, scraping the bowl occasionally; the batter will be thin.

Pour the batter into a greased and floured 9×13-inch cake pan. Bake at 350 degrees for 30 minutes. Cool in pan on a wire rack.

For the frosting, beat the cream cheese and butter in a mixing bowl until creamy. Add the confectioners' sugar, milk and vanilla. Beat until of a spreading consistency, scraping the bowl occasionally. Spread the frosting over the top of the cake.

Serves 15

Chocolate Chip Cake

1 (2-layer) package yellow
cake mix
1 (4-ounce) package vanilla
instant pudding mix
1 cup sour cream
1/2 cup water

1/2 cup vegetable oil
4 eggs
4 ounces German's sweet
chocolate, grated
1 cup semisweet chocolate
chips

Combine the cake mix, pudding mix, sour cream, water, oil and eggs in a mixing bowl. Beat for 4 minutes, scraping the bowl occasionally.

Fold the German's sweet chocolate and chocolate chips into the batter. Spoon the batter into a greased bundt pan. Bake at 350 degrees for 1 hour. Cool in pan for 10 minutes. Invert onto a wire rack to cool completely.

Serves 16

Where if not here? When if not now? Who if not me? That is the philosophy of the Family and Community Education (FCE) program, which is sponsored by the AgCenter and includes more than 2,000 members, mostly homemakers. AgCenter specialists train them in personal development and leadership skills they can pass on in their families and use to better serve Louisiana.

King Cake

2 envelopes dry yeast
2 teaspoons sugar
1/2 cup lukewarm water
31/2 cups flour
1/2 cup sugar
2 teaspoons salt
1 teaspoon nutmeg
1/2 cup lukewarm milk

1 teaspoon grated
  lemon zest
5 egg yolks
1/2 cup (1 stick) butter, cut
  into small pieces
1 cup flour
2 tablespoons butter,
  softened

1 (1-inch) plastic baby,
  dried bean or pecan half
1 egg
1 tablespoon milk
1 tablespoon butter, melted
1/4 cup green-tinted sugar
1/4 cup gold-tinted sugar
1/4 cup purple-tinted sugar

Sprinkle the yeast and 2 teaspoons sugar over the lukewarm water. Let stand until softened; stir. Let stand for 10 minutes longer or until light and bubbly. Combine 31/2 cups flour, 1/2 cup sugar, salt and nutmeg in a mixing bowl and mix well. Stir in the yeast mixture, 1/2 cup lukewarm milk and lemon zest. Beat until blended. Beat in the egg yolks. Add 1/2 cup butter and beat until the butter is blended and the mixture is smooth.

Knead the dough on a lightly floured surface or knead with a mixer dough hook until smooth and elastic, gradually working in 1 cup flour. The dough will not be sticky at the end of the process. Shape the dough into a ball.

Coat the inside of a bowl with 1 tablespoon of the softened butter. Place the dough in the prepared bowl, turning to coat the surface. Let rise, covered with a tea towel, in a draft-free place for 11/2 to 2 hours or until doubled in bulk.

Brush a baking sheet with the remaining 1 tablespoon softened butter. Turn the dough onto a lightly floured surface and shape into a log approximately 14 to 15 inches long. Place the log on the prepared baking sheet and shape into a ring, pressing the ends together to seal. Push a plastic baby, bean or pecan into the cake from the bottom, so that it is not visible from the top.

Let rise, covered with a tea towel, in a warm draft-free place for 45 to 60 minutes or until doubled in bulk. Brush the top with a mixture of the egg and 1 tablespoon milk. Place the baking sheet on the middle oven rack. Bake at 375 degrees for 25 minutes or until brown. Slide the cake onto a wire rack to cool. Brush the top of the cake with 1 tablespoon melted butter. Sprinkle 1/3 of the cake with each color of the tinted sugars.

You may knead candied citron or raisins into the dough, fill the cake with a cinnamon and brown sugar filling or frost with white confectioners' sugar icing before sprinkling with the tinted sugars if desired, but the most common King Cake is plain.

To tint sugar, place a drop of the desired food color into the sugar (one drop per 1/4 cup sugar) and stir until the sugar is evenly colored and brightly tinted.

Serves 15

Chocolate Italian Cream Cake

Cake

2 cups flour, sifted
1/4 cup baking cocoa
1 teaspoon baking soda
1/2 cup (1 stick) butter,
softened

2 cups sugar
1/2 cup shortening
5 egg yolks
1 cup buttermilk

1 cup shredded fresh
coconut
1/2 to 1 cup chopped pecans
1 teaspoon vanilla extract
5 egg whites, stiffly beaten

Chocolate Cream Cheese Frosting

1 (1-pound) package confectioners' sugar
1/4 cup baking cocoa
1/8 teaspoon salt
8 ounces cream cheese, softened

1/2 cup (1 stick) butter, softened
1 teaspoon vanilla extract
1/2 cup chopped pecans

For the cake, mix the flour, baking cocoa and baking soda. Beat the butter, sugar and shortening in a mixing bowl until creamy, scraping the bowl occasionally. Add the egg yolks 1 at a time, beating well after each addition. Add the flour mixture alternately with the buttermilk, beginning and ending with the dry ingredients and beating well after each addition. Stir in the coconut, pecans and vanilla. Fold in the egg whites.

Spoon the batter into 3 greased and floured 8-inch cake pans. Bake at 325 degrees for 25 to 30 minutes or until the layers test done. Cool in pans for 10 minutes. Remove to wire racks to cool completely.

For the frosting, sift the confectioners' sugar, baking cocoa and salt into a bowl and mix well. Beat the cream cheese and butter in a mixing bowl until creamy. Add the vanilla and beat until blended. Add the confectioners' sugar mixture gradually, beating constantly until of a spreading consistency. Spread the frosting between the layers and over the top and side of the cake. Sprinkle the pecans over the top; pat some of the pecans around the side of the cake if desired.

Serves 12

Italian Cream Cake

Cake

1 cup buttermilk
1 teaspoon baking soda
2 cups sugar
1/2 cup shortening
1/2 cup (1 stick) margarine,
  softened

5 egg yolks
2 cups flour
5 egg whites, stiffly beaten
1 cup flaked coconut
1/2 teaspoon vanilla extract
1/2 teaspoon butter flavoring

Nutty Cream Cheese Frosting

8 ounces cream cheese,
  softened
1/2 cup (1 stick) butter or
  margarine, softened

1 (1-pound) package
  confectioners' sugar
1 1/2 cups chopped pecans
1/2 teaspoon vanilla extract

For the cake, mix the buttermilk and baking soda in a small bowl. Combine the sugar, shortening and margarine in a mixing bowl. Beat until creamy, scraping the bowl occasionally. Add the egg yolks 1 at a time, beating well after each addition. Add the buttermilk mixture and flour alternately, ending with the flour and beating well after each addition. Fold in the egg whites. Fold in the coconut and flavorings.

Spoon the batter into 3 floured 9-inch cake pans. Bake at 350 degrees for 25 minutes. Cool in pans for 10 minutes. Remove to a wire rack to cool completely.

For the frosting, beat the cream cheese and butter in a mixing bowl until creamy. Add the confectioners' sugar gradually, beating constantly until of a spreading consistency. Stir in the pecans and vanilla. Spread the frosting between the layers and over the top and side of the cake.

Serves 12

Sour Cream Pound Cake

3 cups sifted flour
1/4 teaspoon baking soda
1/8 teaspoon salt
2 cups sugar
3/4 cup applesauce
1 1/2 cups egg substitute, or 6 eggs
1 cup nonfat sour cream
1 teaspoon vanilla extract
1/2 teaspoon lemon extract
Confectioners' sugar (optional)

Mix the flour, baking soda and salt together. Combine the sugar and applesauce in a mixing bowl and mix well. Add the egg substitute gradually, beating well after each addition. Stir in the sour cream and flavorings. Add the flour mixture gradually and mix until blended.

Spoon the batter into a greased and floured 10-inch tube pan. Bake at 350 degrees for 1 hour and 25 minutes. Cool in pan for 10 minutes. Remove to a wire rack to cool completely. Dust with confectioners' sugar. You may substitute 24 packets or 2 1/2 tablespoons artificial sweetener for the granulated sugar.

Serves 16

Sweet Potato Pound Cake

3 1/2 cups flour
1 teaspoon salt
1 teaspoon baking soda
1 teaspoon baking powder
1 teaspoon each allspice,
cinnamon and nutmeg
3 cups sugar
1 cup shortening
1/2 cup (1 stick) butter or margarine, softened
1 1/2 cups mashed cooked sweet potatoes
6 eggs
1 1/4 cups buttermilk
1/2 teaspoon vanilla extract
1 cup chopped pecans

Mix the flour, salt, baking soda, baking powder, allspice, cinnamon and nutmeg in a bowl. Beat the sugar, shortening and butter in a mixing bowl until creamy, scraping the bowl occasionally. Beat in the sweet potatoes until blended. Add the eggs 1 at a time, beating well after each addition. Add the flour mixture alternately with the buttermilk, beating well after each addition. Beat in the vanilla.

Grease and flour a bundt or tube pan. Sprinkle 1/2 cup of the pecans over the bottom of the prepared pan. Spoon the batter over the pecans and sprinkle with the remaining 1/2 cup pecans. Bake at 350 degrees for 70 minutes. Cool in pan for 10 minutes. Remove to a wire rack to cool completely.

Serves 16

Marshmallow Marble Top Fudge

3 cups milk chocolate chips
1 (14-ounce) can sweetened condensed milk
2 tablespoons butter
1½ teaspoons vanilla extract
1 cup chopped pecans
2 cups miniature marshmallows
2 tablespoons butter

Line an 8×8-inch or 9×9-inch dish with foil. Combine the chocolate chips, condensed milk, 2 tablespoons butter and vanilla in a saucepan. Cook over low heat until blended, stirring frequently. Remove from heat. Stir in the pecans. Spread the chocolate mixture evenly in the prepared dish.

Combine the marshmallows and 2 tablespoons butter in a saucepan. Cook over low heat until blended, stirring frequently. Pour the marshmallow mixture over the prepared layer and swirl with a butter knife for a marbleized effect. Chill for 2 hours or until firm. Cut into squares.

Makes 2 dozen squares

Peanut Butter Fudge

3 cups sugar
½ cup baking cocoa
1 cup milk
2 tablespoons butter or margarine
1 teaspoon vanilla extract
8 ounces peanut butter

Combine the sugar and baking cocoa in a 2-quart saucepan and mix well. Stir in the milk. Cook over low heat until the sugar dissolves, stirring frequently. Bring to a boil. Boil to 234 to 240 degrees on a candy thermometer, soft-ball stage, stirring occasionally. Remove from heat. Add the butter and vanilla; do not stir.

Let stand for 5 minutes. Stir in the peanut butter. Pour into a buttered 8×8-inch dish. Let stand until firm. Cut into squares.

Makes 2 dozen squares

Creamy Pralines

1¹/2 cups packed light
brown sugar
1¹/2 cups sugar
¹/2 teaspoon cream of tartar
¹/8 teaspoon salt

1 cup evaporated milk
¹/2 cup (1 stick) margarine
2 teaspoons vanilla extract
2 cups pecan halves

Place 2 sheets of waxed paper or a sheet of foil over a newspaper on a hard surface. Combine the brown sugar, sugar, cream of tartar and salt in a saucepan and mix well. Stir in the evaporated milk. Bring to a boil. Boil to 240 degrees on a candy thermometer, firm-ball stage, stirring constantly. Remove from heat. Stir in the margarine and vanilla.

Beat for 2 to 3 minutes. Stir in the pecans. Beat until the mixture begins to thicken. Drop by spoonfuls onto the waxed paper. Let stand until firm. If the mixture hardens before completing the dropping process reheat over low heat to melt, or transfer the mixture to a microwave-safe dish and heat in the microwave.

Makes 50 (2-inch) pralines

Nutrition education on demand—that is what you can get through one of the AgCenter's websites called "EatSmart." You can learn about vitamins, minerals, fiber, and how to eat to stay healthy following a "food pyramid" plan. Check our webstite at www.agctr.lsu.edu/eatsmart/fpg.htm—serving Louisiana.

Minted Cheesecake Bars

Crust

1/3 cup butter, melted
1 cup finely crushed chocolate graham crackers

Mint Filling

16 ounces cream cheese, softened
1/2 cup sugar
2 eggs
3/4 teaspoon peppermint extract
1 to 2 drops of green food coloring

Chocolate Glaze

1/4 cup milk chocolate chips
1/2 teaspoon shortening

For the crust, place the butter in a 9×9-inch baking pan. Heat at 350 degrees for 4 to 6 minutes or until melted. Stir in the graham cracker crumbs. Press the crumb mixture over the bottom of the pan.

For the filling, combine the cream cheese, sugar, eggs, flavoring and food coloring in a mixing bowl. Beat for 3 to 4 minutes or until smooth, scraping the bowl occasionally. Spread the filling over the prepared layer. Bake at 350 degrees for 35 to 40 minutes or until set. Cool in pan on a wire rack.

For the glaze, combine the chocolate chips and shortening in a 1-quart saucepan. Cook over low heat until blended, stirring frequently. Drizzle over the cooled baked layer. Chill, covered, until serving time. Cut into bars.

Makes 2 dozen bars

Black Forest Brownies

Brownies

1¹/₃ cups flour
1 teaspoon baking powder
¹/₂ teaspoon salt
1 cup (2 sticks) butter
1 cup baking cocoa
2 cups sugar

4 eggs, beaten
1¹/₂ teaspoons vanilla extract
1 teaspoon almond extract
1 cup maraschino cherries,
    chopped
¹/₂ cup chopped pecans

Chocolate Frosting

2 cups confectioners' sugar
6 tablespoons baking cocoa
¹/₄ cup (¹/₂ stick) butter,
    softened

¹/₄ cup milk
1 teaspoon vanilla extract
¹/₄ cup chopped pecans

For the brownies, mix the flour, baking powder and salt. Heat the butter in a large saucepan until melted. Remove from heat. Stir in the baking cocoa. Add the sugar, eggs and flavorings. Stir in the dry ingredients, cherries and pecans.

Spoon the batter into a greased 9×9-inch baking pan. Bake at 350 degrees for 35 minutes or until the brownies pull from the edges of the pan.

For the frosting, combine the confectioners' sugar and baking cocoa in a mixing bowl and mix well. Add the butter, milk and vanilla. Beat until of a spreading consistency. Spread the frosting over the hot brownies. Sprinkle with the pecans. Let stand until cool. Cut into bars.

Makes 2 dozen bars

Chocolate Chip Cookies

3 cups flour
1 teaspoon baking soda
1 teaspoon salt
1 cup (2 sticks) unsalted butter, softened
1 cup sugar
1 cup packed light brown sugar
2 extra-large eggs
2 teaspoons vanilla extract
3$1/3$ cups semisweet chocolate chips
1$1/2$ cups chopped pecans, toasted

Mix the flour, baking soda and salt. Beat the butter, sugar and brown sugar in a mixing bowl for 3 minutes or until light and fluffy, scraping the bowl occasionally. Add the eggs and vanilla and beat until smooth. Add the dry ingredients gradually, beating constantly until blended. Stir in the chocolate chips and pecans.

Drop the dough by teaspoonfuls 2 inches apart onto a greased cookie sheet. Bake at 325 degrees for 15 minutes or until golden brown. Cool on cookie sheet for 2 minutes. Remove to a wire rack to cool completely. Store in an airtight container.

Makes 5 dozen cookies

Forgotten Cookies

1$1/3$ cups sugar
4 egg whites
2 teaspoons almond extract
1$1/4$ cups butterscotch morsels
1 cup pecan pieces

Beat the sugar and egg whites in a mixing bowl until stiff peaks form. Stir in the flavoring. Fold in the butterscotch morsels and pecans. Overmixing will extract oil from the nuts and butterscotch morsels which in turn will weaken the egg white mixture.

Drop 1 inch apart onto a cookie sheet sprayed with nonstick cooking spray. Place in a preheated 350-degree oven. Turn off the oven immediately. Let stand in the oven with the door closed for 8 to 10 hours. The cookies may be frozen for future use or stored in an airtight container for several days. For variety, substitute chocolate chips for the butterscotch morsels and vanilla extract for the almond extract.

If the cookies stick to the cookie sheet, pass the cookie sheet over a very low flame to release the cookies. Be sure to use pecan pieces because finely chopped pecans release too much oil.

Makes 75 cookies

Mystery Chip Cookies

2 cups (4 sticks) butter, softened
1 cup sugar
3$\frac{1}{2}$ cups flour
2 cups crushed potato chips
2 tablespoons vanilla extract
1 cup white chocolate chips
1 cup chopped macadamia nuts

Beat the butter and sugar in a mixing bowl until creamy, scraping the bowl occasionally. Beat in the flour. Stir in the potato chips and vanilla. Add the chocolate chips and macadamia nuts and mix well.

Drop by teaspoonfuls 2 inches apart onto an ungreased cookie sheet. Bake at 350 degrees for 15 minutes. Cool on cookie sheet for 2 minutes. Remove to a wire rack to cool completely. Store in an airtight container.

Makes 4 dozen cookies

Scrumptious Unbaked Cookies

2 cups sugar
$\frac{1}{2}$ cup 2% milk
$\frac{1}{3}$ cup baking cocoa
$\frac{1}{4}$ cup ($\frac{1}{2}$ stick) margarine
1 cup rolled oats
$\frac{1}{2}$ cup peanut butter
1 teaspoon vanilla extract

Combine the sugar, 2% milk, baking cocoa and margarine in a saucepan. Bring to a boil, stirring occasionally. Boil for 1 minute. Remove from heat. Stir in the oats, peanut butter and vanilla.

Drop by spoonfuls onto waxed paper or foil. Let stand until cool. Store in an airtight container.

Makes 40 cookies

Caramel Ice Cream Pie

Meringue Pie Shell

1 egg white
1/2 teaspoon salt
1/4 cup sugar
1 1/2 cups chopped walnuts

Ice Cream Filling

1 quart coffee ice cream
1 quart vanilla ice cream

Caramel Sauce

2 tablespoons butter
1/2 cup packed brown sugar
1/4 cup light cream
2 tablespoons chopped walnuts
1/2 teaspoon vanilla extract

For the crust, beat the egg white in a small mixing bowl until foamy. Add the salt and beat until soft peaks form. Add the sugar 1 tablespoon at a time, beating constantly until stiff peaks form. Fold in the walnuts.

Spoon the meringue into a buttered 9-inch pie plate with a wet spoon, shaping as for a pie shell. Bake at 400 degrees for 10 minutes. Let stand until cool. Chill in the refrigerator.

For the filling, fill the meringue pie shell with scoops of the coffee and vanilla ice cream. You may cover at this point and freeze until serving time.

For the sauce, heat the butter in a small saucepan. Stir in the brown sugar. Remove from heat. Add the light cream gradually, stirring constantly. Cook for 1 minute, stirring constantly. Pour the sauce into a small heatproof bowl. Stir in the walnuts and vanilla. Drizzle the sauce over the pie just before serving.

Serves 8

Coconut Caramel Pie

1/4 cup (1/2 stick) butter or margarine
1 (7-ounce) package flaked coconut
1/2 cup chopped pecans
1 (14-ounce) can sweetened condensed milk
8 ounces cream cheese, softened
12 to 16 ounces frozen whipped topping, thawed
1 (12-ounce) jar caramel ice cream topping
2 baked (9-inch) pie shells

Heat the butter in a large skillet. Stir in the coconut and pecans. Cook until golden brown, stirring frequently. Remove from heat. Beat the condensed milk and cream cheese in a mixing bowl until smooth, scraping the bowl occasionally. Fold in the whipped topping.

Layer 1/4 of the cream cheese mixture and 1/4 of the ice cream topping in each pie shell. Sprinkle each with 1/4 of the coconut mixture. Repeat the process with the remaining ingredients. Freeze, covered, until firm. Let stand at room temperature for several minutes before serving.

Serves 12

Cushaw Preserves

4 quarts cushaw, peeled, cubed, rinsed
1 1/2 to 2 cups water
6 cups sugar

Combine the cushaw and water in a saucepan and cover. Bring to a boil. Boil just until the cushaw is barely tender. Drain, reserving the liquid.

Combine the reserved liquid with enough additional water to measure 2 cups. Combine with the sugar in a saucepan. Cook over low heat until the sugar dissolves, stirring frequently. Bring to a boil. Stir in the cushaw. Return to a boil over medium heat. Boil for about 1 1/2 hours or until the preserves are clear and the syrup is the consistency of honey, stirring occasionally.

Spoon the preserves into hot sterilized pint jars, leaving a 1/4-inch headspace. Wipe the sealing edge of each jar and seal with 2-piece lids. Process in a simmering water bath for 20 minutes.

Makes 4 pints

Cushaw (kuh-SHAW; KOO-shaw) is a very large, hard-shell squash with a crookneck. It is popular in Cajun and Creole cooking. The meat of the cushaw is yellow and is an excellent source of vitamin A, as well as antioxidants that protect the body from certain types of cancer. It is also high in fiber.

Maw Maw 'Drines Cushaw Pie

Pastry

1 1/2 cups flour	1/2 cup sugar
1 teaspoon baking powder	1 egg, beaten
1/8 teaspoon salt	1 1/2 teaspoons vanilla extract
1/2 cup (1 stick) margarine, softened	2 tablespoons water
	1/2 cup (about) flour

Cushaw Filling

2 cups cushaw or pumpkin preserves (page 199)	2 tablespoons margarine, melted
3 tablespoons milk	1 cup shredded coconut
2 teaspoons vanilla extract	

For the pastry, mix 1 1/2 cups flour, baking powder and salt together. Beat the margarine and sugar in a mixing bowl until creamy. Add the egg and vanilla and mix well. Add the flour mixture alternately with the water, mixing well after each addition. Knead the dough on a lightly floured surface, working in enough of the 1/2 cup flour until the dough is no longer sticky. Reserve some of the pastry for a lattice work design on top. Place the remaining dough in a 9-inch pie plate and with floured hands pat 1/4 inch thick over the bottom and up the side.

For the filling, combine the preserves, milk, vanilla and margarine in a bowl and mix well. Stir in the coconut. Spoon the filling into the pastry-lined pie plate. Roll the reserved pastry on a lightly floured surface and cut into strips. Arrange the strips lattice-fashion over the top of the filling. Bake at 400 degrees for 30 minutes or until the pastry is brown.

Serves 6

Sugar-Free Lemon Ice Box Pie

1 tub sugar-free lemonade mix granules
(enough to make 2 quarts lemonade)
2 (1-ounce) packages sugar-free fat-free vanilla
instant pudding mix
3 1/4 cups skim milk
2 tablespoons nonfat sour cream
Grated zest and juice of 1 medium lemon
12 ounces whipped topping
2 (8-inch) graham cracker pie shells
1 lemon, thinly sliced (optional)

Combine the lemonade mix and pudding mix in a bowl and mix well. Whisk in the skim milk until thickened. Stir in the sour cream, half the lemon zest and lemon juice. Fold in 1/4 of the whipped topping.

Spoon the lemon filling evenly into the pie shells. Chill, covered, in the refrigerator until set. Spread with the remaining whipped topping and sprinkle with the remaining lemon zest. Garnish with lemon slices.

Serves 12

Pecan Pie

1 cup sugar
1 cup light corn syrup
3 eggs, beaten
2 tablespoons butter, chopped into pieces
1 teaspoon vanilla extract
1/4 teaspoon salt
1 cup chopped pecans
1 unbaked (10-inch) deep-dish pie shell
1/2 to 1 cup pecan halves

Combine the sugar, corn syrup, eggs, butter, vanilla and salt in a bowl and mix well. Stir in the chopped pecans. Pour the pecan mixture into the pie shell. Top with the pecan halves.

Arrange the pie plate on a baking sheet. Bake at 300 degrees for 1 hour and 20 minutes. Remove to a wire rack to cool.

Serves 6

More than 250,000 Louisianians suffer from diabetes. Some are not aware of the problem. Others do not get the help they need because of cost. To help prevent unnecessary suffering and death, extension specialists have initiated the Diabetes Education Awareness Recommendations (DEAR) program. Throughout the state, home economists distribute educational and nutritional information about diabetes.

Strawberry Pie

1 teaspoon unflavored gelatin
2 teaspoons water
1 quart sliced fresh strawberries
1 cup sugar
1/4 cup cornstarch
2 tablespoons fresh lemon juice
1 baked (9-inch) pie shell
1/2 cup whipping cream, whipped,
or whipped topping

Soften the gelatin in the water in a small bowl. Mash half of the strawberries in a saucepan. Stir in the sugar and cornstarch. Cook over low heat until thickened and transparent, stirring constantly. Add the gelatin. Cook until dissolved, stirring constantly. Stir in the lemon juice. Let stand until cool.

Fold the remaining strawberries into the thickened filling. Spoon the strawberry filling into the pie shell. Chill until set. Spread with the whipped cream just before serving.

Serves 8

Fresh Strawberry Pie

1 large package sugar-free vanilla pudding mix
1 small package sugar-free strawberry gelatin
3 1/4 cups water
4 cups sliced fresh strawberries
2 packets Sweet 'n Low
2 baked (9-inch) pie shells
8 ounces whipped topping
Strawberries (optional)

Combine the pudding mix and gelatin in a saucepan and mix well. Stir in the water. Bring to a boil over medium heat, stirring occasionally. Remove from heat. Cool slightly.

Toss the sliced strawberries and Sweet 'n Low in a bowl. Spoon the strawberry mixture evenly into the pie shells. Spoon the pudding mixture over the berries. Chill, covered, in the refrigerator until set. Spread with the whipped topping. Garnish with strawberries.

Serves 16

When buying strawberries, look for berries that are fully red. Strawberries do not ripen after being picked. Do not buy bruised or weepy strawberries, and discard any that have molded. Do not rinse until just before use. Besides having a great taste, fresh and frozen strawberries are an excellent source of vitamin C, the water-soluble vitamin being talked about in relation to diet and cancer protection. Strawberries are rich in dietary fiber and contribute iron, riboflavin, and niacin, two of the B vitamins.

Serving Louisiana
Lagniappe

Lagniappe

Lagniappe, *a little something extra,* was always given to customers after a purchase in the stores of early New Orleans. This gave special pleasure to children since normally the lagniappe was a piece of candy. No matter how small the purchase, the merchant always added *something for nothing.*

It seems that the term originated five centuries ago in Normandy and Brittany. Grains like oats, wheat, and barley when sold were spread on a woven cloth known in French as a "nappe." When the seller emptied the contents of the cloth into the buyer's receptacle, there were always a few grains clinging to the cloth. To compensate for this loss, the seller would take one or two handfuls from his stock and throw it into the buyer's bin and say, *pour la nappe,* for the cloth. The custom continued and today in Louisiana that little something extra, lagniappe, is often extended even in a cookbook!

Eating Together

Did you know that most people—children, teens, older adults, and singles—eat more balanced meals and a wider variety of foods when they eat with family or friends? It's true. Mealtimes at the family table help shape and give lasting meaning to our cultural heritage. Positive food memories created during childhood are cherished for life.

Shared meals give us a chance to communicate and build a strong spirit of family or community and commitment to one another. And children learn basic cooking skills and appreciation for a variety of tasty foods when they're involved in meal preparation.

Erma Bombeck wrote about shared family meals. "We argued. We sulked. We laughed. We pitched for favors. We shouted. We listened. It is still our family's finest hour."

Setting the Table

Part of the appeal of any food is its presentation. Not many rules about table settings have changed during the past fifty years. Common sense and consideration for others have always been the basis for placing food and the accompanying dinnerware, flatware, and linens. The rules are easily understood if one examines the "why."

SIMPLE PLACE SETTING

Rule: Place the knife on the right with the cutting edge toward the plate.

Why: When you cut meat, the knife is always in the right hand and the fork in the left, if you are right-handed.

Rule: Place the beverage glass at the tip of the knife.

Why: It is within the natural reach of most people, the majority of whom are right-handed.

Rule: The napkin is usually at the left of the fork with the open corners towards the plate. If space is limited, the napkin may be on the plate.

Why: Very large dinner napkins are awkward to unfold. Putting open corners within easy reach makes it easier to open the napkin and lay it in your lap.

Rule: Place the spoon to the right of the knife.

Why: The spoon is used in the right hand. And art principles suggest that a place setting should have a balanced appearance.

PLACE SETTING WITH SALAD AND BREAD-AND-BUTTER PLATES

Rule: Place the bread-and-butter plate on the left near the top of the fork.

Why: You use your right hand to spread butter, and it is less awkward when the butter plate is to the left.

Rule: The salad plate is usually on the left side at the tip of the fork.

Why: You can reach the salad with your right hand without danger of overturning your beverage.

Rule: When using both salad and bread-and-butter plates, place them to the left as shown above. An alternative is to put the salad plate to the right, below the beverage glass.

Why: Common sense suggests relocating plates to accommodate space. Uniform placement makes dining simple. But right-hand placement can become confusing, especially when a table is set for large crowds with limited space for each place setting.

BUFFET TABLE

Rule: Plates first.
Why: Common sense.

Rule: Main dish next to plate.
Why: This makes serving yourself simpler.
You may place meat on the plate
with two serving utensils, if necessary,
while the plate is still on the table.

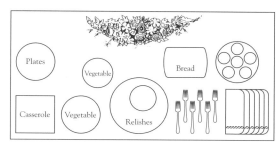

Rule: Vegetables and accompanying food follow the main dish.
Why: It's easier to serve with your right hand while holding the plate in your left.

Rule: Forks and napkins at the end of the food.
Why: The fork may be placed on the plate, and a napkin is easy to hold underneath the plate.

Rule: Beverage last.
Why: Your right hand is now free to pick up and carry the beverage.

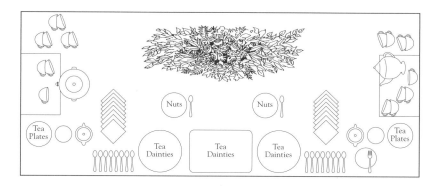

BUFFET TABLE WITH LIGHT REFRESHMENTS

The table with light refreshments for receptions, coffees, dessert parties, and other occasions should be the most attractive area in the room. The illustration above pictures a table with service on one side only. Flowers or other decorations do not have to be limited in height.

The beverage is always served first, followed by napkins, then any necessary flatware and then food. Common sense suggests placing like kinds of food on the same serving tray. Have a duplicate service ready in the kitchen to make it easier to just exchange trays when it is time to refill.

Listed below are a few terms found often when eating one's way through southern Louisiana, courtesy of Louisiana's internationally known Chef John Folse.

Andouille (ahn-do´-ee)—A spicy country sausage used in gumbo and other Cajun dishes.

Beignet (ben´-yea)—Delicious, sweet doughnuts, square-shaped and minus the hole, lavishly sprinkled with powdered sugar.

Bon Appetite! (bon a-pet-tite´)—Literally, good appetite or "Enjoy!"

Boudin (boo´-dan)—Hot, spicy pork mixed with onions, cooked rice, and herbs and stuffed in sausage casing.

Café au Lait (caf-ay´ oh-lay´)—Coffee and chicory blend with milk; usually a half-and-half mixture of hot coffee and hot milk.

Café Brulot (caf-ay´ broo-loh´)—A dramatic after-dinner brew—hot coffee, spices, orange peel, and liqueurs blended in a chafing dish, ignited and served in special cups.

Cajun (cay´-jun)—Slang for Acadians, the French-speaking people who migrated to South Louisiana from Nova Scotia in the 18th century. Cajuns were happily removed from city life, preferring a rustic life along the bayous. The term now applies to the people, the culture, and the cooking.

Chicory (chick´-ory)—An herb, the roots of which are dried, ground and roasted, and used to flavor coffee.

Courtbouillon (coo-boo-yon´)—A rich, spicy soup, or stew, made with fish, tomatoes, onions, and sometimes mixed vegetables.

Crawfish (craw´-fish)—Sometimes spelled "crayfish" but always pronounced crawfish. Resembling tiny lobsters, these little crustaceans are known locally as "mudbugs" because they live in the mud of freshwater bayous. They are served in a variety of ways, including simply boiled.

Creole (cree´-ol)—The word originally described those people of mixed French and Spanish blood who migrated from Europe or were born in Southeast Louisiana. The term has expanded and now embraces a type of cuisine and a style of architecture.

Dirty Rice—Pan-fried leftover cooked rice sautéed with green peppers, onions, celery, stock, liver, giblets, and many other ingredients.

Étouffée (ay-too-fay´)—A succulent, tangy tomato-based sauce. Crawfish Étouffée and Shrimp Étouffée are delicious Louisiana specialties.

Fais do do (fay-doe-doe)—The name for a party where traditional Cajun dance is performed. This phrase literally means "to make sleep," although the parties are the liveliest of occasions.

Filé (fee´-lay)—Ground sassafras leaves used to season, among other things, gumbo.

Grillades (gree´-yads)—Squares of braised beef or veal. Grillades and grits is a popular local breakfast.

Gumbo (gum´-boe)—A delicacy of South Louisiana. A thick, robust soup with thousands of variations, only a few of which are Shrimp Gumbo, Chicken Gumbo, Okra Gumbo, and Filé Gumbo.

Jambalaya (jum´-bo-lie´-yah)—Another many-splendored thing. Louisiana chefs "sweep up the kitchen" and toss just about everything into the pot. Tomatoes and cooked rice, plus ham, shrimp, chicken, celery, onions, and a whole shelf full of seasonings.

King Cake—A ring-shaped pastry decorated with colored sugar in the traditional Mardi Gras colors—purple, green, and gold, that represent justice, faith, and power. A small plastic baby is hidden inside the cake and the person who finds the baby must provide the next King Cake.

Lagniappe (lan´-yap)—This word is Cajun for "something extra," like the extra doughnut in a baker's dozen. An unexpected nice surprise.

Mirliton (mel-e-taun´)—A hard-shelled squash, sometimes called a vegetable pear, with edible innards. It is cooked like squash and stuffed with either ham or shrimp and spicy dressing.

Muffuletta (muf-a-lot´-ta)—This huge sandwich is made up of thick layers of several different types of Italian meats, cheeses, and a layer of olive salad. Served on special Muffuletta bread.

Praline (praw´-leen)—The sweetest of sweets, this New Orleans tradition is a candy patty. The essential ingredients are sugar, water, and pecans.

Po-Boy—Another sandwich extravaganza, which began as a 5-cent lunch for—who else?—poor boys. There are fried oyster po-boys, roast beef and gravy po-boys, softshell crab po-boys, and others, all served upon crispy-crusted loaf bread called French Bread.

Red Beans and Rice—Kidney beans cooked in seasonings and spices and usually with big chunks of sausage and ham—served over a bed of rice.

Sauté—To cook in a skillet containing a small amount of hot cooking oil. Sautéed foods should never be immersed in the oil. Should be stirred frequently.

Menu Items

Beverages

Coffee

Punch or lemonade

Tea, iced, 12-ounce glass

Tea, hot

Half-and-half, cream for coffee

Lemons for tea

Sugar for tea, coffee

Meat

Beef, lamb, roasted, sliced

Ham, bone in, sliced

Ham, boneless, sliced

Ham, ground

Turkey, sliced

Turkey for casserole or spaghetti

Hamburger

Meat loaf

Pork rib roast

Pork chops and veal chops

Chicken, whole, cooked, and cubed for casseroles

Chicken for salad

Chicken, creamed

To Serve 50 You Will Need

1^1/2 pounds for 75 cups

2^1/2 gallons

6 (1-ounce) family-size tea bags and
3 gallons water

^{1}/2 pound tea for 75 cups

1^1/2 pints

6 cut into 8 wedges or 8 slices

^{3}/4 pound

16 to 18 pounds

22 pounds

15 pounds

12 pounds

2 (20-pound) turkeys

1 (12-pound) turkey

18 pounds

10 pounds

20 pounds

18 pounds

6 (4-pound) chickens

4 quarts, cubed (see above) with
4 quarts celery and 2 quarts mayonnaise

6 quarts chicken cubed and 3^1/2 quarts
white sauce

Menu Items

Salad

Fruit salad	7 quarts prepared fruit, drained
Gelatin	10 envelopes gelatin, 3 quarts liquid and 2$1/2$ quarts solid ingredients
Lettuce	4 heads for leaf under salad and 7 heads for a tossed salad
Salad dressing	1 quart
Mayonnaise for "dollops"	1 quart (1$1/2$ tablespoons per serving)
Cabbage (for slaw)	5 pounds

Desserts

Ice cream	2 gallons
Salted nuts	3 to 4 pounds
Whipping cream for garnish	1 quart (1$1/2$ tablespoons per serving)
Cakes	3 or 4 (9-inch layer or 10-inch angel food), 14 to 16 slices each

Soups — 4 gallons

Dairy Products

Butter for table	1 pound (72 pats per pound)
Butter for sandwiches	1 pound for 100 slices bread
Butter for vegetable seasoning	$1/2$ pound; $1/2$ teaspoon per serving
Cheese for sandwiches or with cold cuts	5 pounds

Miscellaneous

Rolls	7 dozen
Dip	4 to 6 cups
Ice	2 to 4 large bags

To Serve 50 You Will Need

Protein

Why?   Absolutely essential in building, repairing, and renewing of all body tissue. Helps body resist infection. Builds enzymes and hormones, helps form and maintain body fluids.

Where?   Milk, eggs, lean meats, poultry, fish, soy beans, peanuts, dried peas and beans, grains, and cereals.

Carbohydrates

Why?   Provide needed energy for bodily functions, provide warmth, as well as fuel for brain and nerve tissues. Lack of carbohydrates will cause body to use protein for energy rather than for repair and building.

Where?   Sugars: sugar, table syrups, jellies, jams, etc. as well as dried and fresh fruits. Starches: cereals, pasta, rice, corn, dried beans and peas, potatoes, stem and leafy vegetables, and milk.

Fats

Why?   Essential in the use of fat-soluble vitamins (A, D, E, K) and fatty acids. Have more than twice the concentrated energy than equal amount of carbohydrate for body energy and warmth.

Where?   Margarine, butter, cooking oil, mayonnaise, vegetable shortening, milk, cream, ice cream, cheese, meat, fish, eggs, poultry, chocolate, coconut, nuts.

Vitamin A

Why?   Needed for growth, healthy skin, bones, and teeth. Helps maintain good vision, especially in dim light. Helps body resist infection.

Where?   Fish-liver oils, liver, butter, cream, whole and fortified milk, whole-milk cheeses, egg yolk, dark green leafy vegetables, and fortified products.

Vitamin D

Why?   Helps to maintain concentration of calcium and phosphorus in the blood, which aids in their absorption and use to promote healthy bones and teeth.

Where?   Fish-liver oils, fortified milk, exposure to sunlight.

Vitamin E

Why?   Helps retard destruction of vitamin A and ascorbic aid. Protects red blood cells.

Where?   Widely distributed among many foods. Cereal seed oils, such as wheat germ, soybeans, corn, and cottonseed.

Vitamin K

Why? Helps promote normal blood clotting.

Where? Green leaves such as spinach and cabbage. Cauliflower and liver.

Vitamin C

Why? Formation of collagen (a material that holds cells together). For healthy teeth, gums, and blood vessels. Aids in healing wounds and resisting infection.

Where? Citrus fruits, tomatoes, strawberries, cantaloupe, cabbage, broccoli, kale, potatoes.

Thiamin

Why? Promotes the use of carbohydrates for energy and helps maintain healthy nervous system.

Where? Pork, liver and other organs, brewer's yeast, wheat germ, whole-grain or enriched cereals and breads, soybeans, peanuts and other legumes, and milk

Riboflavin

Why? Helps use nutrients for energy and tissue building. Promotes healthy skin and eyes.

Where? Milk, organ meats, and enriched breads and cereals.

Niacin

Why? Needed for healthy nervous system, skin, and normal digestion. Helps cells use oxygen to release energy.

Where? Lean meat, fish, poultry, liver, kidney, whole wheat and enriched cereals and breads, peanuts and brewer's yeast

Folate

Why? Necessary for development of red blood cells and normal metabolism of nutrients. Can prevent neural tube defects in babies if consumed by women of childbearing age.

Where? Liver, legumes, yeast, and deep green leafy vegetables are highest sources.

B-6

Why? Needed for use of protein. Prevents certain forms of anemia.

Where? Wheat germ, meat, liver, kidney, whole-grain cereals, soybeans, peanuts, and corn.

B-12

Why? Helps prevent certain forms of anemia. Needed for proper growth and a healthy nervous system.

Where? Liver, meat, milk, eggs, and cheese.

Biotin
Why?   Needed for normal metabolism of carbohydrates, protein, and fats.
Where?   Cauliflower, organ meats, egg yolk, and legumes.

Pantothenic Acid
Why?   Helps to metabolize nutrients to produce energy. Aids in the synthesis of amino acids, fatty acids, and hormones.
Where?   Almost universally present in plant and animal tissue. Especially rich sources: liver, yeast, eggs, peanuts, and whole-grain cereals.

Choline
Why?   Needed for normal function of cells and cell membranes.
Where?   Lettuce, peanuts, coffee, and cauliflower.

Calcium
Why?   Assists in clotting of blood and building bones and teeth. Minimizes bone loss and can help prevent osteoporosis. Promotes proper function of nerves, heart, and muscles.
Where?   Milk, cheese, ice cream, turnips, collards and mustard greens, broccoli, and cabbage.

Phosphorus
Why?   Needed for bones and teeth and for enzymes used in energy metabolism. Acts as blood buffer.
Where?   Milk, cheese, ice cream, meat, poultry, fish, whole-grain cereals, nuts, and legumes.

Magnesium
Why?   Needed for regulation of body temperature, contraction of nerves and muscles, and synthesis of protein. A cofactor for enzymes involved in cellular metabolism.
Where?   Nuts, soybeans, seafood, whole grains, meat, dried peas and beans.

Iron
Why?   Makes hemoglobin, the red substance in blood. Transports oxygen to and from cells.
Where?   Organ meats, oysters, meats, leafy green vegetables, dried peas, enriched breads and cereals.

Iodine
Why?   Needed for regulation of the use of energy in body. Prevents goiter.
Where?   Seafood, iodized salt.

Zinc
Why? Needed for wound healing, normal development, healthy skin, and many body chemical reactions.
Where? Meat, liver, eggs, shellfish, green leafy vegetables, fruit.

Copper
Why? Needed for proper use of iron, red blood cell formation, and part of many enzymes.
Where? Oysters, nuts, liver, kidney, dried legumes.

Manganese
Why? Part of several enzymes. Minor component in bone.
Where? Nuts and unrefined grains are rich sources. Vegetables and fruits contain moderate amounts.

Fluoride
Why? For normal formation and development of teeth and bones. Decreases tooth decay.
Where? Fluoridated water.

Chromium
Why? Maintaining normal glucose metabolism.
Where? Brewer's yeast, meat products, cheeses, whole grains.

Selenium
Why? Part of enzyme that protects cells against oxidation.
Where? Seafood, kidney, liver, meat. Grains vary, depending on where they are grown.

Potassium
Why? Involved in fluid balance and muscle activity.
Where? Potatoes, prunes, oranges, bananas, red meat, and whole-grain products.

Sodium
Why? Maintenance of fluid balances of acids and bases in the body.
Where? Sodium chloride (table salt), most processed and canned foods.

Chloride
Why? Maintenance of fluid and electrolyte balance. Component of gastric juice.
Where? Sodium chloride (table salt).

The editors have attempted to present these family recipes in a format that allows approximate nutritional values to be computed. Persons with dietary or health problems or whose diets require close monitoring should not rely solely on the nutritional information provided. They should consult their physician or a registered dietitian for specific information.

Abbreviations for Nutritional Profile

Cal — Calories	T Fat — Total Fat	Sod — Sodium
Prot — Protein	Chol — Cholesterol	g — grams
Carbo — Carbohydrates	Fiber — Dietary Fiber	mg — milligrams

Nutritional information for these recipes is computed from information derived from many sources, including materials supplied by the United States Department of Agriculture, computer databanks, and journals in which the information is assumed to be in the public domain. However, many specialty items, new products, and processed food may not be available from these sources or may vary from the average values used in these profiles. More information on new and/or specific products may be obtained by reading the nutrient labels. Unless otherwise specified, the nutritional profile of these recipes is based on all measurements being level.

- Artificial sweeteners vary in use and strength and should be used to taste, using the recipe ingredients as a guideline. Sweeteners using aspartame (NutraSweet® and Equal®) should not be used as a sweetener in recipes involving prolonged heating, which reduces the sweet taste. For further information on the use of these sweeteners, refer to the package.
- Alcoholic ingredients have been analyzed for the basic information. Cooking causes the evaporation of alcohol, which decreases alcoholic and caloric content.
- Buttermilk, sour cream, and yogurt are the types available commercially.
- Canned beans and vegetables have been analyzed with the canning liquid. Rinsing and draining canned products will lower the sodium content.
- Chicken, cooked for boning and chopping, has been roasted; this method yields the lowest caloric values.
- Eggs are all large. To avoid raw eggs that may carry salmonella, as in eggnog or 6-week muffin batter, use an equivalent amount of commercial egg substitute.
- Flour is unsifted all-purpose flour.
- Garnishes, serving suggestions, and other optional information are not included in the profile.
- Margarine and butter are regular, not whipped or presoftened.
- Oil is any type of vegetable cooking oil. Shortening is hydrogenated vegetable shortening.
- Ingredients to taste have not been included in the nutritional profile.
- If a choice of ingredients has been given, the profile reflects the first option. If a choice of amounts has been given, the profile reflects the greater amount.

Pg #	Recipe Title (Approx Per Serving)	Cal	Prot (g)	Carbo (g)	T Fat (g)	% Cal from Fat	Chol (mg)	Fiber (g)	Sod (mg)
13	Creole Turtle Soup[1]	537	30	43	27	45	109	6	1970
14	Swiss Steak Royale	545	59	38	17	28	141	4	2027
15	Harry D. Wilson's Hash	643	43	68	23	31	90	6	1050
16	Oven-Baked Roast	488	50	8	27	51	166	1	912
16	Meat, Noodle and Cheese Casserole	972	47	81	52	48	135	5	1095
17	Bill's Bobotie	570	37	40	29	46	177	3	928
18	Pork Chops à la Vivi	290	34	19	8	26	86	1	253
19	Chicken Jambalaya	801	42	78	34	39	112	2	698
20	Ro-Tel Chicken	503	41	44	18	32	102	4	1233
20	Chicken Enchilada Casserole	546	25	29	38	61	87	3	1265
21	Delicious Barbecued Shrimp	343	15	1	31	81	135	<1	754
22	Blackened Fish Fillets	415	51	<1	22	49	207	0	789
23	The Howard[2]	1048	90	6	73	63	310	<1	2606
24	Nick's Baked Beans	288	20	29	11	33	54	4	468
24	Green Bean Casserole	201	7	11	15	65	21	2	1192
25	Sweet Potato Bake	842	8	118	40	41	74	5	556
26	Bread Pudding with Rum Sauce	445	7	68	16	33	121	1	255
28	Phillips Phamily Phavorite	309	3	42	15	43	52	1	309
31	Avocado and Shrimp Mash	36	1	3	2	56	10	1	57
31	Crab Canapés	77	3	4	6	66	10	<1	225
32	Chunky Avocado Dip	233	3	10	22	80	9	8	347
33	Best Crab Dip	68	3	1	6	75	20	<1	101
33	Hot Crab Dip	85	3	5	6	62	24	2	203
34	Crawfish Dip	81	4	3	6	66	46	<1	132
34	Hamburg Dip	107	3	2	10	83	32	<1	236
35	Mexican Dip	395	8	6	38	86	47	3	473
35	Mexicorn Dip	153	4	6	13	74	22	1	327
36	Strawberry Dip	201	1	36	6	28	18	1	102
36	Monterey Jack Salsa	119	3	4	10	76	10	1	388
37	Spinach and Artichoke Dip	139	5	4	12	76	33	2	176
37	Dried Beef Cheese Ball	140	6	2	12	79	42	<1	525
38	Armadillo Eggs	45	2	5	2	42	11	<1	138
39	Bourbon Bites	152	4	8	10	59	21	<1	424
39	Picante Chicken	117	11	1	8	59	46	<1	180
40	Turkey Tarts	109	3	6	8	69	13	<1	113
41	Salmon Tarts	89	2	7	6	60	24	<1	104
41	Crab and Cream Cheese Bake	156	5	8	12	67	46	<1	229
42	Sausage and Cheese Squares	62	3	2	4	66	26	<1	131

Pg #	Recipe Title (Approx Per Serving)	Cal	Prot (g)	Carbo (g)	T Fat (g)	% Cal from Fat	Chol (mg)	Fiber (g)	Sod (mg)
42	Sugared Pecans	585	5	58	41	59	0	5	1
43	Bourbon Slush	158	1	32	<1	0	0	<1	2
43	Cocktail Slush	124	1	23	<1	1	0	<1	6
44	Mimosa	100	<1	13	<1	1	0	<1	5
44	Tiger Punch	57	<1	14	<1	1	0	<1	1
45	Golden Punch	51	<1	13	<1	1	0	<1	3
45	Rosé Sparkle	110	<1	9	<1	0	0	1	12
46	Orange Breakfast Shake	192	5	38	3	13	10	2	67
46	Hot Spiced Tea	142	<1	38	0	0	0	<1	2
49	Chili	302	23	28	12	34	54	9	1556
50	Mémère JuJu's Broccoli Soup	409	24	20	26	57	70	3	1636
51	Cabbage and Sausage Soup	489	19	68	20	34	18	10	2256
51	Cheese Soup	412	23	22	25	56	72	4	2657
52	Chicken Noodle Soup	192	14	27	3	15	29	3	174
52	Corn and Crab Soup	Nutritional profile for this recipe is not available.							
53	Cream of Crawfish Soup	613	16	16	55	80	260	1	490
53	French Potato Soup	233	8	19	15	55	8	2	752
54	Tortilla Soup	111	4	18	3	26	1	3	501
54	Vegetable Beef Soup	215	12	29	6	24	27	5	634
55	Cajun Chicken and Sausage Gumbo[3]	477	17	36	29	55	94	2	815
56	Seafood Filé Gumbo	291	29	27	6	21	242	1	315
57	Shrimp and Okra Gumbo	308	20	41	7	21	174	5	1573
58	Cranberry Salad	298	4	51	11	31	0	3	82
58	Marinated Fruit Salad	137	1	34	<1	2	0	3	13
59	Apricot Salad	258	3	38	11	37	17	1	72
59	Southern Pretzel Strawberry Salad	348	3	39	20	51	17	1	260
60	Nine-Day Coleslaw	259	2	24	19	62	0	3	1190
60	Herbed Tomato Salad	129	1	5	12	83	0	1	203
61	Bleu Cheese Chicken Salad	141	15	5	7	45	37	2	253
62	Polynesian Chicken Salad	316	17	12	22	64	57	2	362
63	Mexican Fiesta Chicken and Rice Salad	339	20	51	6	16	33	8	1000
64	Seafood Rice Salad	441	10	25	32	67	188	1	371
64	Shrimp and Pasta Salad[3]	368	18	43	13	32	111	2	626
67	Rib-Eye Roast with Oven-Browned Vegetables	681	45	31	41	55	139	3	411
68	Slow-Cooker Roast	487	49	7	28	52	164	<1	823
68	Baked Brisket	323	40	8	14	39	124	1	96
69	Beef and Beer Buffet	248	36	7	6	21	94	<1	173
70	Chinese Pepper Steak	323	28	18	16	43	75	4	1443

Pg #	Recipe Title (Approx Per Serving)	Cal	Prot (g)	Carbo (g)	T Fat (g)	% Cal from Fat	Chol (mg)	Fiber (g)	Sod (mg)
71	Rosemary Pepper Beef Steaks	337	48	1	15	41	139	<1	103
71	Slow-Cooker Beef Stew	343	32	28	11	28	94	4	105
72	Beer Stew	644	35	53	29	40	94	5	600
73	Stuffed Bell Peppers	259	12	33	8	28	27	5	264
74	Eggplant-Stuffed Bell Peppers	530	32	49	24	41	81	10	199
75	Lasagna	446	31	18	28	56	144	2	1228
76	No-Boil Easy Lasagna	658	42	37	38	52	126	5	1362
77	Creole Meatballs	372	21	19	23	56	91	2	837
78	The Kids Will Never Know Poorboys	627	29	55	32	46	77	4	1401
79	Neapolitan Beef Pie	1234	74	56	77	57	238	3	2956
79	South-of-the-Border Bake	568	32	32	34	55	98	1	1534
80	"More" Casserole	711	44	74	28	35	165	5	935
81	Roast Pork	472	59	6	22	43	172	<1	840
81	Pork and Cabbage Casserole	502	26	35	29	51	97	4	512
82	Savory Ham Pie	Nutritional profile for this recipe is not available.							
83	Chicken Spaghetti	513	28	38	27	48	67	2	698
84	Chicken and Corn Bread Dressing	Nutritional profile for this recipe is not available.							
85	Old-Fashioned Chicken and Dumplings	645	36	52	31	45	93	2	1003
86	Easy Chicken and Dumplings	411	29	34	17	37	67	2	668
86	Chicken and Gravy[4]	415	39	5	26	56	119	1	386
87	Chicken Pie	768	36	38	52	61	122	1	981
87	Oven-Barbecued Chicken	238	28	24	4	13	73	1	2248
88	New Orleans Chicken	406	29	2	31	69	89	5	273
88	Chicken and Crawfish	333	33	16	15	40	189	3	799
89	Easy Chicken Quesadillas	415	24	44	15	33	50	3	732
89	Baked Chicken Parmesan	392	33	25	17	39	77	1	534
90	Italian Chicken	204	29	1	9	39	93	<1	486
91	Stuffed Italian Chicken Rolls	300	36	12	11	34	99	<1	417
92	Chicken and Angel Hair Pasta	288	19	35	8	24	27	2	378
93	Quick Chicken Jambalaya	481	16	18	38	71	135	1	396
93	Chicken Corn Bread Bake	224	14	25	7	30	47	2	797
94	Chicken Enchiladas	269	12	16	18	59	55	2	585
94	Barbecue Sauce	85	<1	4	8	82	0	<1	830
95	Fried Turkey	Nutritional profile for this recipe is not available.							
95	Pinto Beans with Smoked Turkey	283	28	36	3	9	67	14	483
96	Stuffed Cornish Game Hen[5]	1504	130	75	75	45	548	8	471
97	Venison Stew	408	36	19	20	45	122	1	81
97	Venison and Squash	297	14	12	21	64	112	2	985

Pg #	Recipe Title (Approx Per Serving)	Cal	Prot (g)	Carbo (g)	T Fat (g)	% Cal from Fat	Chol (mg)	Fiber (g)	Sod (mg)
98	Venison Pot Roast	416	52	23	12	27	191	<1	242
101	Catfish Co-Op	458	34	51	12	24	87	6	847
102	Catfish à la Creole	293	25	13	15	45	77	3	254
103	Crunchy Catfish with Lemon Parsley Sauce	642	40	8	50	70	304	1	1367
104	Oven-Fried Catfish	441	38	35	16	33	116	3	161
104	Gaspergou with Clear Sauce	490	69	6	19	37	248	1	294
105	Salmon Balls[6]	164	22	11	4	21	99	1	422
105	Baked Salmon Croquettes	192	16	14	8	39	57	<1	544
106	Parmesan Flounder Fillets	404	27	<1	32	72	100	<1	366
107	Deviled Crab	273	7	12	22	72	192	2	1907
108	Stuffed Crabs	215	6	18	13	55	39	1	440
109	Boiled Crabs[1]	252	47	5	4	15	228	1	6834
110	Boiled Crawfish[7]	1829	317	84	23	11	2413	12	37437
110	Crawfish Easy	588	26	62	26	40	184	3	1086
111	Crawfish Étouffée	348	18	15	25	62	183	3	1442
112	South Louisiana Crawfish Fettuccini	813	44	43	53	58	240	3	1575
113	Blue Ribbon Crawfish Quiche	428	18	21	30	64	236	1	332
114	Crawfish Supreme	602	16	21	50	73	92	2	1639
115	Asparagus and Shrimp Oriental	78	5	5	5	55	20	2	906
116	Cocodrie Baked Shrimp[8]	902	28	43	65	66	265	2	2279
117	Herbed Shrimp and Feta Cheese Casserole	249	22	19	9	34	114	<1	784
118	Shrimp My Way	275	31	9	13	43	300	3	862
118	Sopping Good Shrimp	415	29	<1	32	70	352	0	624
119	Boiled Shrimp[7]	242	29	2	12	47	269	<1	8683
119	Shrimp in Beer	276	30	6	12	39	269	1	2064
120	Shrimp and Crab Linguini	360	34	23	15	37	197	1	961
121	Shrimp and Crab-Stuffed Peppers	845	50	67	42	44	411	7	1623
122	Shrimp Casserole	413	17	40	21	45	150	2	980
123	Shrimp Fettuccini	682	26	36	48	63	292	2	758
124	Shrimp Étouffée	252	20	6	16	59	221	1	364
124	Barbecued Shrimp Kabobs	287	15	20	17	52	149	3	1132
125	Shrimp Pie	280	14	16	18	57	87	1	402
126	Shrimp with Rice	431	27	35	17	36	257	2	907
127	Seafood au Gratin	316	26	12	18	52	244	1	823
128	Seafood Potato Casserole	230	16	23	9	33	84	2	215
129	Rice Cooker Seafood Jambalaya	467	30	43	19	36	219	2	932
130	Louisiana Baked Oysters	206	9	14	12	54	82	1	374
130	Cocktail Sauce	250	1	9	24	85	20	1	448

Pg #	Recipe Title (Approx Per Serving)	Cal	Prot (g)	Carbo (g)	T Fat (g)	% Cal from Fat	Chol (mg)	Fiber (g)	Sod (mg)
131	Oysters à la LuLa	274	10	30	10	34	46	1	257
132	Alligator Sauce Piquant	441	45	26	15	31	0	3	470
135	Asparagus Casserole	167	7	21	7	36	6	6	1087
136	Red Beans and Rice	233	19	36	2	7	19	8	569
136	Ginger Beets	330	3	80	2	5	4	4	100
137	Broccoli and Rice Casserole	462	14	40	28	54	23	4	1096
138	Glazed Carrots	218	2	38	8	31	21	5	142
138	Corn Pudding	336	8	46	16	40	166	2	599
139	Louisiana Corn Creole	452	24	35	25	49	126	3	891
140	Eggplant with Shrimp[9]	355	31	31	12	30	271	6	551
141	Stuffed Mirliton	245	17	22	10	37	45	3	647
142	Okra Croquettes[6]	71	3	12	2	19	53	2	313
142	Potatoes Supreme	384	10	22	29	68	75	1	744
143	Sweet Potato Casserole with Pecan Topping	250	4	51	4	16	<1	4	115
144	Southern-Style Yellow Squash	77	2	9	4	45	10	4	43
144	Squash Casserole	164	8	15	9	45	18	3	536
145	Fried Green Tomatoes[10]	1066	4	18	111	92	21	1	334
145	Zucchini Rice Bake	301	20	27	12	36	110	2	250
146	Festive Holiday Yams	331	2	67	7	17	12	4	104
147	Vegetable Batter	53	1	7	2	37	13	<1	295
147	Hot Sherried Fruit	254	1	48	8	27	21	2	96
148	Garlic Grits	346	12	20	24	63	39	<1	1137
148	Crawfish Corn Bread Dressing	Nutritional profile for this recipe is not available.							
149	Mexican Corn Bread Casserole	460	24	35	25	49	106	3	930
149	Cajun Eggplant Dressing	197	6	33	5	21	30	3	1269
150	Rice Casserole	145	3	31	1	7	1	2	282
150	Green Tomato Pickles	15	<1	4	<1	2	0	<1	126
153	Sweet Potato Biscuits	211	3	24	12	50	1	1	445
153	Cowboy Coffee Cake	358	4	57	13	33	38	1	265
154	Cranberry Coffee Cake	574	6	69	32	48	114	2	571
155	Gingerbread	163	3	36	1	6	25	<1	156
155	Beer Bread	136	3	28	<1	2	0	1	520
156	Old-Fashioned Louisiana Corn Bread	103	2	13	4	39	19	1	517
156	Raisin Bran Muffins	191	4	25	10	43	18	3	197
157	Lemon Bread	293	4	35	16	48	58	1	159
158	Lemon Blueberry Bread	223	3	37	7	28	52	1	140
159	Anadama Bread	187	5	34	4	18	18	1	301
160	No-Knead Maple Molasses Bran Bread	76	2	14	2	20	7	2	73

Pg #	Recipe Title (Approx Per Serving)	Cal	Prot (g)	Carbo (g)	T Fat (g)	% Cal from Fat	Chol (mg)	Fiber (g)	Sod (mg)
161	Classic Cloverleaf Yeast Rolls	176	3	24	7	38	14	1	82
162	Refrigerator Rolls	66	2	12	1	18	4	<1	82
163	Whole Wheat Refrigerator Rolls	Nutritional profile for this recipe is not available.							
164	Flensjes[11]	820	29	137	16	18	254	3	226
165	Overnight French Toast	191	6	27	7	33	50	1	321
165	Make-Ahead Breakfast Bake	307	17	3	25	73	360	0	517
166	Eggs Fantastic	642	40	9	49	69	348	1	1468
167	Brunch Casserole	347	17	13	24	64	155	0	814
167	Cajun Eggs[3]	310	18	4	24	71	264	<1	850
168	Corny Egg Casserole	365	13	29	23	56	152	2	695
168	Breakfast Pizza	342	15	18	23	61	146	<1	870
171	Bananas Foster	531	3	73	23	38	70	2	225
171	Chocolate Turtle Cheesecake	638	8	81	33	45	28	3	509
172	Favorite Peach Cobbler	514	4	61	29	50	2	3	505
173	Bread Pudding with Bourbon Sauce	287	9	39	7	23	121	1	257
174	Death by Chocolate	334	4	41	16	42	36	1	394
174	Guiltless Peach Dessert	211	10	38	3	11	3	1	455
175	Pumpkin Roll	204	3	27	9	41	67	1	173
176	Grandmother's Favorite Butter Tarts	265	3	35	14	45	23	1	83
177	Raisin Tarts	276	2	30	17	55	56	1	138
178	Fruit Sherbet	186	1	48	<1	1	0	2	6
178	Cantaloupe Sorbet	98	1	25	0	0	0	1	13
179	Melon Ice	132	2	33	<1	2	0	1	17
180	Vanilla Ice Cream	Nutritional profile for this recipe is not available.							
181	Apple Fig Cake	276	3	46	10	30	21	2	206
182	Applesauce Cake	502	6	71	23	40	58	4	358
183	Blueberry Cake	273	4	29	16	52	55	<1	55
184	Carrot Cake with Cream Cheese Frosting	835	7	92	51	53	112	2	598
185	Carrot Cake with Orange Glaze	270	4	64	<1	1	0	1	247
186	Chocolate Cake	490	4	72	22	39	46	1	330
187	Chocolate Chip Cake	361	4	44	20	49	60	1	329
188	King Cake	311	6	47	11	32	109	1	408
189	Chocolate Italian Cream Cake	805	9	94	46	50	152	3	391
190	Italian Cream Cake	801	9	94	46	50	131	2	392
191	Sour Cream Pound Cake	215	6	46	1	4	<1	1	91
191	Sweet Potato Pound Cake	528	7	68	26	44	96	2	362
192	Marshamallow Marble Top Fudge	225	3	26	13	51	15	1	60
192	Peanut Butter Fudge	173	3	28	6	31	4	1	59

Pg #	Recipe Title (Approx Per Serving)	Cal	Prot (g)	Carbo (g)	T Fat (g)	% Cal from Fat	Chol (mg)	Fiber (g)	Sod (mg)
193	Creamy Pralines	101	1	14	5	45	2	<1	35
194	Minted Cheesecake Bars	211	3	20	13	57	46	<1	158
195	Black Forest Brownies	277	3	38	14	43	62	1	179
196	Chocolate Chip Cookies	151	2	19	9	49	17	1	65
196	Forgotten Cookies	41	<1	6	2	42	0	<1	5
197	Mystery Chip Cookies	167	2	15	11	61	21	1	96
197	Scrumptious Unbaked Cookies	81	1	13	3	33	<1	1	30
198	Caramel Ice Cream Pie	687	12	62	45	58	165	2	298
199	Coconut Caramel Pie	656	8	67	39	54	43	2	452
199	Cushaw Preserves[12]	630	2	160	1	1	0	5	7
200	Maw Maw 'Drines Cushaw Pie	698	7	110	26	33	36	3	410
201	Sugar-Free Lemon Ice Box Pie	320	4	41	15	43	1	1	459
201	Pecan Pie	789	9	94	44	49	117	4	394
202	Strawberry Pie	311	3	46	14	39	20	2	132
202	Fresh Strawberry Pie	187	2	20	10	51	0	2	253

[1]Nutritional profile includes all of the ingredients.

[2]Nutritional profile reflects the use of seasoned salt.

[3]Nutritional profile reflects the substitution of seasoned salt for Cajun seasoned salt.

[4]Nutritional profile includes the entire amount of the flour mixture.

[5]Nutritional profile does not include mushroom sauce.

[6]Nutritional profile does not include vegetable oil for deep-frying.

[7]Nutritional profile includes all of the ingredients except liquid crab boil.

[8]Nutritional profile reflects the substitution of clam juice for reduced shrimp stock.

[9]Nutritional profile includes the entire amount of milk.

[10]Nutritional profile includes the entire amount of vegetable oil.

[11]Nutritional profile does not include the vegetable oil.

[12]Nutritional profile reflects the substitution of crookneck squash for cushaw.

Serving Louisiana

LSU Agricultural Center
P.O. Box 25203 • Baton Rouge, Louisiana 70894-5203
phone: 225-578-4161 • fax: 225-578-4143

Name

Street Address   Telephone

City   State   Zip

Your Order	Quantity	Total
Serving Louisiana at $19.95 per book		$
EBRP residents add sales tax at $1.80 per book Other Louisiana residents add $ .80 per book		$
Shipping and handling at $3.00 per book		$
Total		$

Method of Payment: [   ] American Express   [   ] Discover   [   ] MasterCard   [   ] VISA
[   ] Check enclosed payable to LSU Foundation.

Account Number   Expiration Date

Signature

For the lastest research-based information on just about anything, visit our website at www.lsuagcenter.com. For a catalog of FREE publications, call 225-578-2263.

Photocopies accepted.